Ouha YASSINE

Arganière Reserve: a natural climatic bulwark

Ouha YASSINE

Arganière Reserve: a natural climatic bulwark

Imprint

Any brand names and product names mentioned in this book are subject to trademark, brand or patent protection and are trademarks or registered trademarks of their respective holders. The use of brand names, product names, common names, trade names, product descriptions etc. even without a particular marking in this work is in no way to be construed to mean that such names may be regarded as unrestricted in respect of trademark and brand protection legislation and could thus be used by anyone.

Cover image: www.ingimage.com

This book is a translation from the original published under ISBN 978-620-6-70459-1.

Publisher:
Sciencia Scripts
is a trademark of
Dodo Books Indian Ocean Ltd. and OmniScriptum S.R.L publishing group

120 High Road, East Finchley, London, N2 9ED, United Kingdom
Str. Armeneasca 28/1, office 1, Chisinau MD-2012, Republic of Moldova, Europe
Managing Directors: Ieva Konstantinova, Victoria Ursu
info@omniscriptum.com

Printed at: see last page
ISBN: 978-620-3-13146-8

Contribution to the assessment of the role of the Arganier biosphere reserve in mitigating the effects of climate change through carbon storage (Central zone of the RBA)

Summary

Recent attempts to mitigate the effects of climate change have focused on carbon sequestration by forests because of their potential to absorb CO_2 from the atmosphere. However, the consequences of actual forest management practices on carbon storage capacity are still controversial. With this in mind, our study focused on the effect of argan banking (ABR) on the carbon sequestration potential of argan forest ecosystems in the biogeographical zone of the Souss plain and Dir.

Two sites were studied (Admine and Ouameslakht), and three zones were identified for each: the core zone, the buffer zone and the transition zone. Analysis of carbon sequestration potential at the Admine site showed that there was no significant difference between the 3 zones, while the Ouameslakht site showed a significant effect between them

In the Admine site, the total carbon stocks of the different zones are respectively: 65; 50.73 and 31.63 t C ha^{-1} for the central, buffer and transition zones. As for the Ouameslakht site, the total carbon stocks of the different zones are respectively: 44.75; 31.84 and 19.66 t C ha^{-1} for the central, buffer and transition zones.

Analysis of land-use dynamics between 1998 and 2021, in the two core zones, has shown that in the Admine core zone, the argan forest has regressed in favor of rangeland as a result of strong anthropic pressure, despite the presumed restriction of all human activity, and so the objective assigned to this zone has not yet been achieved.

In the central Ouameslakht zone, the forest has undergone a reduction in area in favor of less productive land, making it relatively well preserved compared with the central Admine zone, thanks to the introduction of fencing in this area.

Keywords: climate change, carbon sequestration, ASR, core areas, land use dynamics, Admine, Ouameslakht

Contents

Introduction

Climate change is a phenomenon that is currently the subject of much discussion throughout the international scientific community, with the 5th report of the Intergovernmental Panel on Climate Change (IPCC) confirming that climate change is having a significant impact on the "land sector" (agriculture, forestry, land use change). In 2018, the IPCC's special report recommended that it would be necessary to achieve carbon neutrality on a global scale by 2050. This ambitious objective of balance between anthropogenic greenhouse gas (GHG) emissions and CO_2 sequestration by ecosystems is now the benchmark for most international climate policies (Sylvain Pellerin et al., 2019).

For this reason, the Kyoto Protocol (1997) called for the assessment and reporting of national emissions of atmospheric carbon reflected by changes in carbon reservoirs, and for the establishment of a precise inventory of carbon stocks and accumulation potential in forests. In addition, Ryan (1991) considered that knowledge of carbon fluxes through vegetation is of paramount importance in understanding the current state and predicting future evolution of forest ecosystems. Carbon flows through forests are closely linked to their dynamism and vigour (Campagna, 1996). Similarly, the amount of organic carbon in a forest soil is the result of the balance between net vegetation production and organic matter decomposition (Liski and Westman, 1997).

However, Mediterranean forests have been relatively little studied from this point of view, due in particular to their low commercial wood productivity, which ignores their role in mitigating GHG emissions (Shaiek et al., 2010).

Among the Mediterranean forests are those of the argan tree (*Argania spinosa*), a species endemic to south-west Morocco, which extends over an area of 868,034 ha (IFN, 1999). Since 08/12/1998, the argan grove has been declared Morocco's first Biosphere Reserve by UNESCO, covering an area of over 2.5 million hectares.

First of all, the main vocation of a Biosphere Reserve is to achieve the lasting conservation of an entire ecosystem in order to maintain sustainability. The

Arganeraie, with all its natural and human complexities, extends over several biogeographical units (mountains, plains, wetlands and drylands). To structure and balance the reserve's functions, those in charge have zoned it into several interdependent units: 18 core areas (Zone A) covering 16,620 hectares, 13 buffer zones (Zone B) covering 582,450 hectares and transition zones or "Zone C", comprising the areas of the RBA not covered by Zones A and B (DEF, 2020).

However, at present, there is no information on the effects of biosphere reserve designation on the contribution of argan forests and their ecosystems to carbon sequestration and therefore to climate change mitigation, which is why accurate and effective quantification of carbon stocks in the stands and soils of argan ecosystems in the different land-use zones of the RBA would be crucial to understanding their response to climate change mitigation.

Based mainly on biophysical criteria, the RBA has been subdivided into three main agroecological zones, each with its own specificities in terms of land use patterns, biodiversity resources and ecosystem services. The following three zones were therefore taken into consideration: the High Atlas zone; the Plain zone and the Anti-Atlas zone.

With this in mind, the aim of the present work is to estimate the response of carbon stocks to management methods, in particular the establishment of the reserve, between 1998 and 2021 in the Central zone of the RBA (Plaine du Souss and Dir).

To achieve this goal, we propose :

- Quantify and evaluate the carbon stocks of the various occupations in 2021 according to the zones of the ASR (core, buffer, transition);

- Mapping land use in the study area in 1998 and 2021;

- Analyze the spatio-temporal monitoring of land use dynamics in the central areas of the Souss plain and Dir : Admine and Ouameslakht.

Part 1: Bibliographical summary

Chapter 1: Argan tree ecology

1.1. Taxonomy and botanical description of the argan tree

The argan tree (Argania spinosa) belongs to the order Ebenales and the family Sapotaceae. This tropical family comprises around 600 species in some 50 genera (Emberger, 1960).

Resembling an olive tree, the argan is described as a third-growth tree (Boudy, 1950), reaching 8 to 10 m in height depending on the ecological conditions of the environment. The crown is very large and spread out. The trunk is short (2 to 3 m), gnarled, twisted and often even multiple, made up of several intertwined stems. The branches are dense with spiny tips, hence the name "spinosa", on which abundant fruiting bodies bloom.

Leaves are often fasciculate, whole, lanceolate or spatulate, more or less petiolate. They are dark green on the upper surface, lighter underneath. Foliage is evergreen. However, in the event of severe or prolonged drought, argan trees are forced to shed their leaves to resist evaporation, and start budding and budding again, sometimes several weeks before the rainy season resumes (Emberger, 1960).

The argan flower is hermaphroditic (Boudy, 1952). The calyx and corolla are made up of five sepals and five petals respectively. The androecium consists of five stamens with short threads. The ovoid ovary, topped by a conical style, contains only two or three uniovulated carpels. The fruit of the argan tree, the morphological study of which has been the subject of numerous works (Emberger, 1960), is a sessile berry consisting of a fleshy pericarp or pulp and a "pseudo-endocarp" or stone, containing the seeds, which are generally fused together and vary in number from one to several per stone. According to shape and size, six types of fruit have been distinguished: fusiform, oval, apiculate oval, drop, rounded, globular.

The roots of the argan tree are highly developed, and can be tracing when hard rock opposes its extension, enabling it to take advantage of even small amounts of rain.

1.2. Distribution of the argan tree in Morocco

The argan tree is a plant species endemic to Morocco. It occupies an area of 868,034 hectares (IFN, 1999). It forms steppe formations in south-western Morocco between the Oued Tensift and Oued Noun; it extends from Safi to Sidi Ifni, in the Souss plain and on the western flanks of the Western High Atlas and the Anti Atlas (Benabid, 2000).

In the north of the country, it is found in isolated stands in two particular stations: the Oued Grou near Rabat and the Beni Snassen massif west of Berkane (Maire, 1939), spread over some 800 ha.

In terms of altitude, the species ranges from sea level up into the mountains, reaching 1500 meters in the warmest regions (the mountains bordering the Souss plain). This limit drops slightly towards the north.

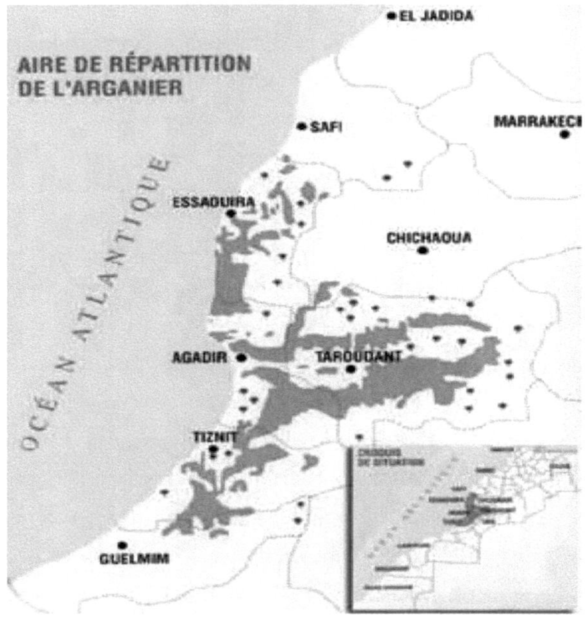

(Source: Peltier in M'hirit et al., 1998)

Figure1 : Map of argan distribution in southwest Morocco

1.3. Argan tree ecology

The argan tree is a thermophilous and xerophilous species, but it requires a mild climate in general, with high air humidity on the Atlantic coast, although the oceanic influence diminishes as we move further inland. In the argan area, the climate is Mediterranean (Emberger, 1955).

The sub-Mediterranean is the vegetation stage of the argan grove (Achhal, 1986). Its bioclimate corresponds to the arid stage for two-thirds of its area, while the rest 8 on the coast from Safi to Agadir is a transition zone towards the semi-arid (Peltier, 1982).

From an edaphic point of view, this species is virtually indifferent to soil type once climatic conditions have been met. It is highly plastic with respect to soil, thriving on schist, limestone and alluvium. However, mobile sands are not hospitable to the tree, as the root system becomes exposed and dries out. (Rieuf, 1962).

Chapter 2. Overview of MAB and biosphere reserves

2.1.MAB_UNESCO program

The Man and the Biosphere Programme (known as MAB) was launched in 1971 by Unesco. MAB is an intergovernmental scientific programme aimed at establishing a scientific basis for improving the relationship between people and their environment. It combines the natural and social sciences to improve people's livelihoods and safeguard natural and managed ecosystems, promoting innovative approaches to economic development that are socially and culturally appropriate and environmentally sustainable. This is why the biosphere reserve concept was originally developed in 1974, and was substantially revised in 1995 with the adoption by UNESCO's General Conference of the Statutory Framework and the Seville Strategy for Biosphere Reserves. Their guidelines, adopted in 1996, define the operating principles of biosphere reserves.

Figure 1 shows the MAB Council's perception of biosphere reserve zones:

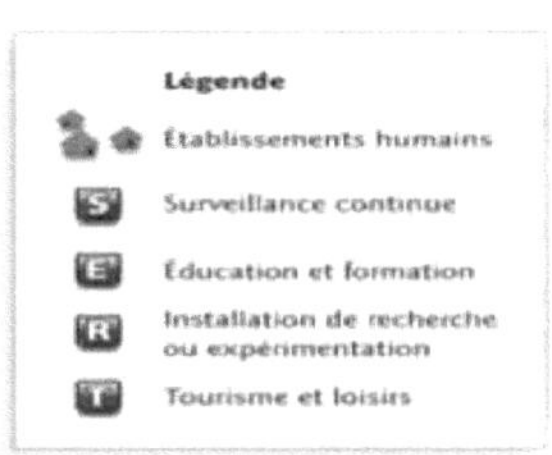

(Source: UNESCO, 2000)

Figure2 : Biosphere reserve zoning

2.2.Protected areas and biosphere reserves in Morocco
2.2.1. Protected areas in Morocco

Renowned for its rich biodiversity and high-quality landscapes and natural environments, Morocco is one of the countries most committed to preserving its natural heritage, and its pioneering experience with protected areas in particular makes it a model to follow.

This particular interest in nature conservation illustrates the Kingdom's commitment to a policy of sustainable development, which aims both to safeguard its biological diversity and to protect endangered species.

Morocco has been involved in the creation of protected areas since the 1930s, when it promulgated a dahir in 1934 enabling the creation of national parks. The purpose of these protected areas is to conserve, enhance and rehabilitate natural and cultural heritage, conduct scientific research, raise public awareness and provide entertainment, promote ecotourism and contribute to sustainable economic and social development.

Morocco has ten national parks to its credit. These include Toubkal National Park in 1942, Tazekka National Park in 1950, Sous Massa National Park in 1991, Iriqi National Park in 1994, Ifrane, Talassamtane, Haut Atlas Oriental and Al Hoceima National Parks in 2004, Khenifiss National Park in 2006, the Kingdom's first Saharan national park, and Khenifra National Park in 2008, in addition to the extension of Tazekka National Park in 2004 and Ifrane National Park in 2008.

2.2.2. Biosphere reserves in Morocco

In addition to this network of national parks, Morocco has four Biosphere Reserves, which promote solutions that reconcile the conservation of biodiversity with its sustainable use. These are :

2.2.2.1.The Arganeraie Biosphere Reserve (ABR)

The RBA is the first biosphere reserve to be created in 1998 in south-west Morocco. The reserve was created around an endemic Moroccan forest species, the Argan tree (Argania spinosa), which has great biogeographical value and is the main feature of the Moroccan Macaronesian sector. The argan forest provides multiple functions and uses for populations whose socio-economic activities are strongly linked to the various products provided by the argan grove. The Arganeraie Biosphere Reserve (ABR) covers a vast, intra-monotonous plain with its mountainous edges, all of which is open to the Atlantic Ocean to the west.

2.2.2.2. The Biosphere Reserve of the Oases of Southern Morocco (RBOSM)

In November 2000, the area of the three southern Moroccan provinces of Ouarzazate, Errachidia and Zagora was recognized by UNESCO as the "Southern Moroccan Oases Biosphere Reserve". With the recognition of the biosphere reserve status, this region has become an integral part of UNESCO's Man and the Biosphere (MAB) World Program.

Traditional resource management systems go hand in hand with social and cultural structures based on active solidarity in the development of infrastructures, particularly for the exploitation and mobilization of water resources (Khettaras).

A global strategy for safeguarding the Oases was then required. It mobilized the various players involved in regional development, technical services, universities and research institutes, while taking into account the opinions of all stakeholders. The Ministry of Agriculture, Maritime Fishing, Rural Development and Water and Forests, in partnership with the Moroccan MAB National Committee, is in charge of managing and supervising the project.

2.2.2.3. The Mediterranean Intercontinental Biosphere Reserve (MIMBR)

The RBIM is a unique Biosphere Reserve. Like many other reserves, it is transboundary, but the only intercontinental one in the world. It covers an area of almost one million hectares, and is shared almost equally between the Moroccan and Spanish shores. The Moroccan part is located in the heart of the Tingitane peninsula. For the most part, it concerns only the mountainous part of the Western Rif, known as Jbala country. It covers a large part of the Province of Chefchaouen and variable areas of the Wilaya of Tétouan and the Provinces of Fnidek, Fahs-Anjra and Larache.

On the Moroccan side, the RBIM contains numerous natural ecosystems (forest, coastal and marine) of great bio-ecological value. Some of these have been selected as Sites of Biological and Ecological Interest (SIBE), identified in this area of Morocco by the study of protected areas carried out between 1993 and 1995: the Talassemtane National Park site, Jbel Bouhachem, Jbel Moussa, Smir Lagoon, Oued Tahaddart...

The aim of the RBIM is to improve environmental conditions and work towards sustainable development, while trying to create and consolidate channels of communication and participation for local communities, and develop cooperation between the two shores. The Moroccan part of the RBIM is intended to provide an environmentally balanced framework for a region currently undergoing economic expansion.

2.2.2.4.Atlas Cedar Biosphere Reserve (RBCA)

The Cedar Biosphere Reserve, located in the center of the Atlas Mountains, was recognized by UNESCO as a World Heritage Site in 2016. This area is home to the majestic Atlas cedar, which accounts for almost %75 of the tree's global population.

Covering a large part of the Atlas Mountains, the cedar forest, a major bastion of Berber culture, provides the region with vitally important water resources, as well as areas for grazing and agriculture, and potential for tourist activities, which, following on from an exclusive semi-nomadic pastoral tradition, exert strong pressure on natural resources.

In addition to natural forest ecosystems, the reserve includes a mosaic of medium-altitude open rangeland areas characteristic of the landscapes and Amazigh lifestyles and cultures of the Middle Atlas. The RBCA's status enables it to achieve its conservation and development objectives, by capitalizing on projects to manage and protect the forest massifs of the Ifrane and Khénifra provinces, notably in the Ifrane, Haut Atlas Oriental and Khénifra National Parks.

2.3.The Arganeraie Biosphere Reserve
2.3.1. History of the RBA

Following the worrying situation of the degradation of the argan grove, caused mainly by strong anthropic pressure, the State has implemented the recognition of this area as an Arganeraie Biosphere Reserve (ABR) of the UNESCO Man and Biosphere (MAB) program, by applying the recommendations of the Seville strategy drawn up by the MAB committee assembly held in Seville (Spain) in March 1995.

On December 8, 1998, the Arganeraie was declared Morocco's first Biosphere Reserve by UNESCO.

2.3.2. Geographical location

The RBA's intervention zone is located in the biogeographical area where the argan tree grows. The RBA covers an area of around 2.5 million hectares. It covers the provinces and prefectures of Agadir Ida Ou Tanane, Inezgane Ait Melloul, Chtouka Ait Baha, Tiznit, Taroudant, Sidi Ifni and Essaouira.

2.3.3. Definition and objectives

UNESCO's 1998 designation of the Arganeraie as a biosphere reserve set the following objectives:

> ➢ Establish an overall diagnosis, including that relating to the phytoecological and biodiversity context (Bendaanoun, 1998);
> ➢ Achieve sustainable conservation of the entire argan ecosystem, to maintain the sustainability or at least the continuity of the processes and mechanisms of the ecosystem's natural evolution;
>
> ➢ Conservation, enhancement and regeneration of argan tree stands to ensure sustainable development;
> ➢ Reasoned and rational use of resources and local know-how;

2.3.4. RBA functions
2.3.4.1. Ecosystem conservation function

This function concerns the protection of biodiversity, which is the major challenge and goal of biosphere reserve designation. A biosphere reserve is an environment where a representative sample of the ecosystem concerned can be conserved in a sustainable manner, thereby maintaining biodiversity and promoting the continuity of the natural functioning and evolution of natural ecosystems.

2.3.4.2. Training and awareness function

To ensure the conservation and sustainable development of the argan ecosystem, it is essential to develop training and educational activities. The RBA provides theoretical and practical training centers for researchers, the local population and visitors alike.

2.3.4.3.Enhancement and development cooperation function

The natural resources generated by the RBA need to be developed, while preserving the long-term regeneration potential of the natural ecosystem. In order to reconcile sustainable development with rational use to meet the needs of the local population, the organization of the latter into cooperatives and associations seems an essential necessity.

2.3.5. The RBA 2002 master plan

With the aim of implementing RBA, a Framework Plan was drawn up between 1998 and 2002 as part of the Projet Conservation et Développement de l'Arganeraie (PCDA), co-financed by German cooperation through GTZ. A Network of Local Development Associations involved in RBA (RARBA) was also set up in 2002.

The reserve has been zoned according to the three zones stipulated in the UNESCO MAB network standards for the creation of Biosphere Reserves (Figures 3 and 4):

Core areas or long-term integral protection zones, to conserve biological diversity, monitor the least disturbed ecosystems and conduct scientific research. The choice of areas was based on the land covered by the Sites of Biological and Ecological Interest (SIBE), including the Souss-Massa National Park (PNSM). In addition, other zones were proposed by local managers, in consultation with local communities and civil society (communes, NGOs). As a result, 18 core zones have been identified in the RBA, with a combined surface area of at least 17,000 ha, ranging from 300 ha (Ait Er-Rkha) to 4,200 ha (Souss Massa National Park). They correspond to the SIBEs identified by the Moroccan Protected Areas Master Plan and to the sites proposed by the managers:

> ➢ Either the ecological context linked to the presence of the argan tree and the difficult access.
> ➢ Or the existence of a significant and interesting natural phenomenon, such as the presence of a well-established stand of argan trees, a rare animal or plant species.
> ➢ The absence of human activity (houses, Azibs, tracks, etc.).

Buffer zones surrounding or juxtaposed to the core areas and are intended to be managed with a view to production compatible with ecologically sustainable practices.

In fact, 13 buffer zones have been identified, with a combined surface area of around 560,000 ha. Their selection was based on the following criteria:

> Existence and importance of argan trees (scattered or low-density stands are excluded, land at risk of erosion is included).
> Importance of the argan tree in the local economy.

Transitional zones, comprising areas of the RBA not covered by zones A and B. The objective assigned to these zones is to achieve sustainable socio-economic development of the Arganeraie area. These flexible transition zones may contain a number of agricultural activities, human settlements or other exploitations, and in which local communities, management agencies, scientists, NGOs, economic and cultural interests and other partners work together to manage and sustainably develop the region's resources in a spirit of solidarity with other zones and especially with the core zones. . In the RBA, Zone C also includes towns and cities (Greater Agadir, Essaouira, Taroudant, Tiznit).

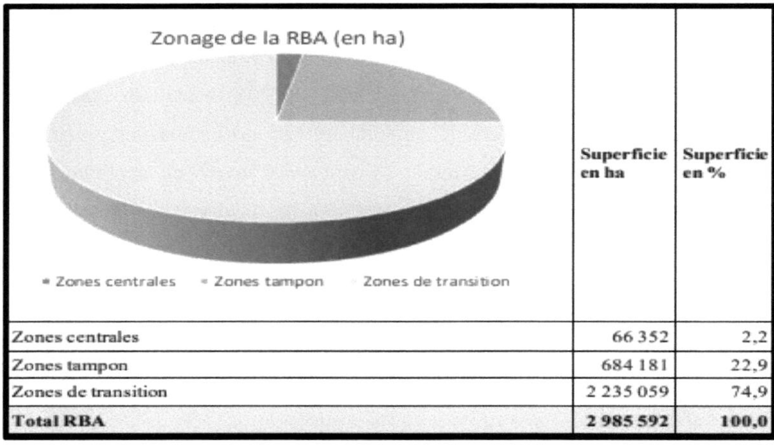

	Superficie en ha	Superficie en %
Zones centrales	66 352	2,2
Zones tampon	684 181	22,9
Zones de transition	2 235 059	74,9
Total RBA	**2 985 592**	**100,0**

(Source: DEF,2020)

Figure3 : Importance of RBA zones

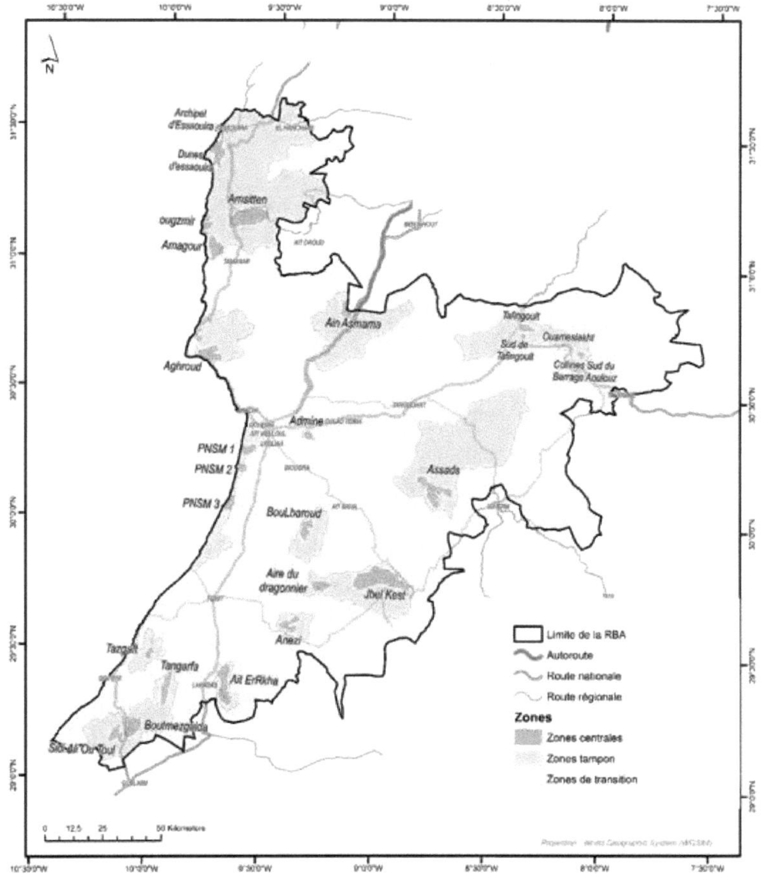

(Source: DEF,2020)

Figure4 : RBA zoning map

Chapter 3: Climate change

3.1.On a global scale

Climate change can be defined as a variation in the state of the climate, which can be detected by changes in the mean and/or variability of its properties, and which persists over an extended period, usually decades or more. Climate change may be due to natural internal processes or external forcings, including modulations of solar cycles, volcanic eruptions or persistent anthropogenic changes in atmospheric composition or land use (IPCC, 2007).

According to the IPCC Special Report, the global average temperature has risen by around 1°C (IPCC, 2018). A number of studies have shown that, in addition to the + 1.5°C warming, there is a significant risk of exceeding the climate tipping point "tipping points", which can lead to uncontrolled climate change (Moore, 2018).

Although it is difficult to identify precisely when this tipping point may be passed, it is certain that the next few decades will be crucial (IPCC, 2018).

Indeed, compared with pre-industrial levels, scenarios that limit global warming to +1.5°C require rapid, far-reaching and unprecedented changes in all aspects of our social organization. The report estimates that the atmosphere cannot absorb more than 420 gigatonnes (Gt) of CO_2 to stay at this temperature. Below the threshold, humans emit around 42 Gt of CO_2 globally every year, which means that at the current rate, this limit can be exceeded in 9 years, and there are still 26 years to go before the upper limit of global warming, +2°C, can be exceeded. It should be possible to achieve this goal by reaching "net zero emissions" by 2050, it should be possible to maintain the +2°C target.

As a result, several articles in the IPCC (2018) report emphasize the importance of natural ecosystems, including forests, as they absorb CO_2 and keep the rising global average temperature (below 2°C) as close as possible to 1.5°C.

3.2.Nationwide

Morocco's meteorological and geographical position makes it a vulnerable region to climate change, both in terms of temperature and rainfall.

Thermal indices (temperature and rainfall) show a warming trend that is more pronounced in the eastern part of the country, on the mountainous relief and piedmont. Rainfall indices show a drying trend, especially at the end of the rainy season - an important period for agriculture - and a tendency for the semi-arid climate to migrate northwards. Due to its subtropical latitude, Saharan influences and Mediterranean conditions, Morocco is likely to experience higher rates of warming than other regions of the world (Mhirit and Et-Tobi, 2010).

The fifth IPCC report predicts a decrease in rainfall for Morocco of up to 20% by the end of the century (IPCC, 2014).

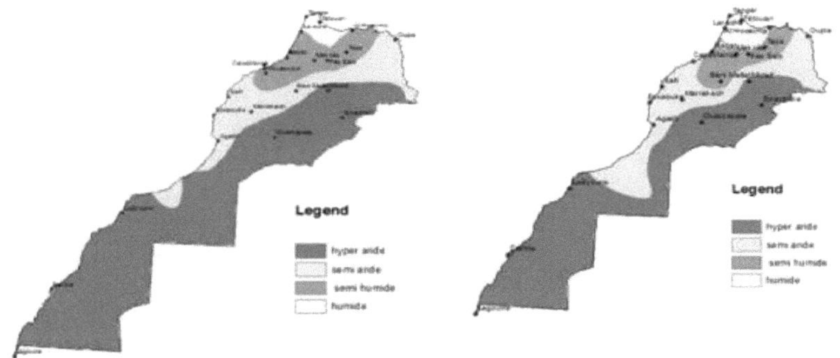

(Source: rainfall trends from 1960 to 2015 in Morocco)

Figure5 : Map showing the spatial distribution of climate types according to the Martonne index over the period 1980-1990 (Left) and 1990-2000 (Right).

Regions that were classified as humid and sub-humid climates are regressing in favor of semi-arid and arid climates; following the increase in average annual temperature by 0.16°C per decade and the 47% drop in spring precipitation nationwide (Figure 5) (Mokssit, 2012).

3.3.Protected areas: a major asset in the fight against climate change

Global warming is creating new challenges for the sustainable management of natural resources in protected areas. In particular, this is due to the fact that protected areas are a spatially static management tool in the face of a spatially dynamic problem (climate variability, species dispersal and adaptation). This problem can be solved in part by more effective, adaptive management of protected areas. However, all this leads us to examine the capacity of protected areas as an important vector in the fight against climate change (Halpin, 1997; Heller and Zavaleta, 2009). If effectively managed, they can play an important role in both adaptation and mitigation.

Chapter4. Carbon stocks

4.1.   Global carbon cycle

Although it is not the most abundant chemical element on Earth, carbon is one of the most important compounds in the functioning and evolution of the Earth system. A series of complex biogeochemical processes enable its transfer from one reservoir to another, leading to a planetary cycle. (Figure 2)

The latter is divided between a fast cycle, involving the atmosphere, ocean and biosphere reservoirs, and a slow cycle, involving the earth's crust, soils and ocean. As carbon dioxide is a greenhouse gas, the carbon cycle interacts very closely with the climate machine, resulting in the establishment of feedback loops that regulate or, on the contrary, speed up its functioning. These loops are the key elements in the functioning of the cycle.

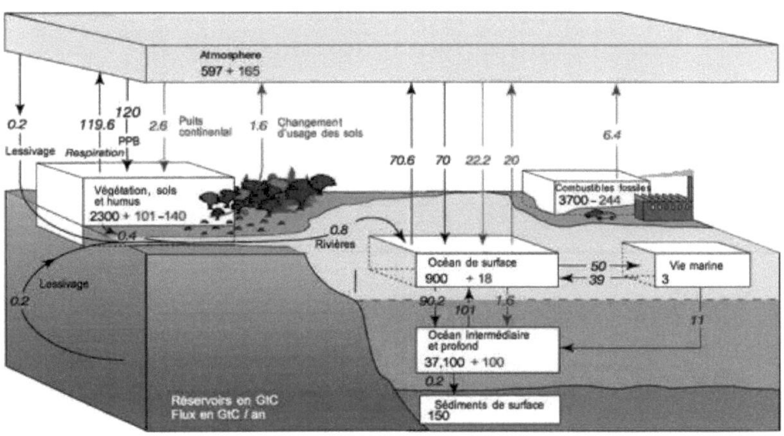

(Source: Sarmiento, 2002)

Figure6 : Global carbon cycle

4.2.   Protected areas and carbon stocks

Protected areas cover almost 15.3% of the world's land surface, including inland waters (Maxwell et al., 2020), but their contribution to combating climate change remains insufficiently understood. They contribute notably to optimizing carbon sequestration and storage by avoiding deforestation and land and forest cover degradation (Noumi et al., 2018).

By combating deforestation and land degradation, protected areas help maintain carbon stocks and capture, as well as contribute to climate equilibrium (Lewis et al., 2009). These protected areas were designed primarily to protect biodiversity from direct human impact, but they can also be used to combat climate change, over and above their primary role of protecting ecosystems.

Take, for example, the contribution of protected areas to the carbon stock of Central African countries (Figure 3). Protected areas cover around 17.6% of the land area of Central African countries, and contain between 20% and 25% of the carbon stock of these countries.

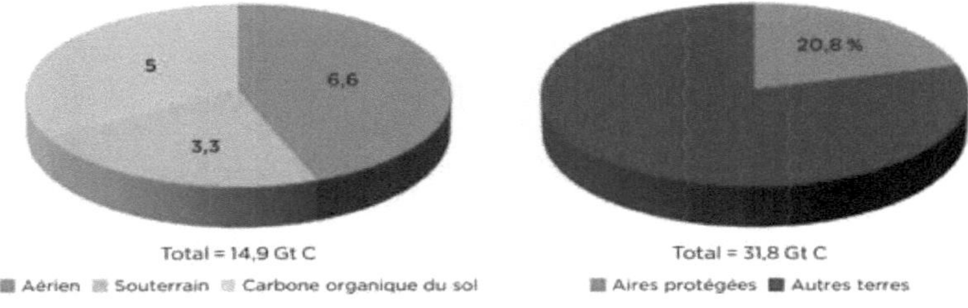

Figure7 : Carbon stocks in the Central African protected area network

4.3.   The role of forests in the carbon cycle

On a global scale, forests contain around half of the carbon accumulated by terrestrial ecosystems, while covering just under 30% of the earth's surface. Forest photosynthesis recycles 5% of atmospheric CO_2 annually: without forests, it is estimated, with considerable uncertainty, that the rate of increase in the greenhouse effect and associated climate change would be 1.5 times faster.

The dynamics of terrestrial ecosystems depend on the interactions of several biogeochemical cycles, including the carbon cycle, nutrient cycles and the water cycle. Forests play a key role in the carbon cycle, operating dynamically to recycle carbon. The processes of photosynthesis, respiration, transpiration, decomposition and combustion maintain the natural circulation of carbon between ecosystems and the atmosphere.

Human activities can modify carbon stocks and exchanges. Significant carbon emissions have resulted from deforestation over the last few centuries in mid- and high latitudes and, in the latter part of the 20th century, in tropical regions. Today, tropical deforestation is responsible for 20% of global annual greenhouse gas (GHG) emissions. However, on the whole, forests act as carbon sinks, absorbing more carbon than they emit. On a global scale, the annual balance is around 1 billion tonnes of carbon sequestered by forest ecosystems, taking deforestation into account (Marianne, 2009).

4.4. Carbon stock in plant matter

The most important processes involving carbon in the terrestrial ecosystem are: photosynthesis, respiration, translocation, allocation, storage, fine root turnover, decomposition, the influence of herbivores and leaf or other shedding (Landsberg et al., 1991). Photosynthesis, respiration and allocation, among others, are independent and vary according to environmental conditions, ecosystems and stand age (Ryan, 1991). Despite the difficulty of measuring these processes, it is nevertheless possible to estimate carbon fluxes from annual changes in tree biomass (Kozlowski et al., 1991).

- **Gross primary production (GPP) :**

This is the total mass of organic compounds produced by photosynthesis.

- **Net primary production (NPP) :**

This is the difference between SCH and autotrophic respiration. It represents the mass of organic matter synthesized by plants. NPP therefore includes all increases in the mass of stems, leaves, reproductive organs, roots and the amount of plant tissue consumed by herbivores or which dies and becomes detritus (Waring and Schlesinger, 1985).

- **Biomass :**

The total mass of living organisms within a given perimeter or volume at the time of observation. For some time now, dead plants have often been included in biomass.

- **Net ecosystem production (NEP) :**

This is the difference between PPB and autotrophic and heterotrophic respiration. This concept excludes other carbon exports from the ecosystem, such as those caused by fire, logging, erosion or others (Waring and Schlesinger, 1985). Evaluating different ecosystems around the world using variables such as PPN enables us to assess the contribution of forest ecosystems to the global carbon cycle. Waring and Schlesinger (1985) reveal that forests and forest soils are excellent carbon reservoirs, and outperform all other terrestrial ecosystems in terms of NPP, with the exception of marshes and swamps.

4.5.   Soil carbon stock

Soil is a biogeochemical material with biological and structural properties. It is a mechanical support for plants and a source of the nutrients, water and air needed for their growth. It fulfils a range of environmental, forestry and agronomic functions. Soil contributes to the carbon cycle by storing carbon and releasing it into the atmosphere.

Soil organic carbon (SOC) is the main component of soil organic matter (SOM) and, as such, is the fuel of all soils. SOM is one of the key functions of soil, as it is essential for stabilizing soil structure, and retaining and releasing nutrients from plants. It enables infiltration and storage of water in the soil. It is therefore essential for soil health, fertility and food production. SOC loss indicates a certain level of soil degradation.

In the global carbon cycle, forest soils are gigantic carbon reservoirs (Arrouays et al., 2002), and can act as carbon sinks or sources, depending on how the land is used and managed, and on environmental changes. In fact, it is estimated that they store around 80 tC / ha in the first 30 centimetres, which concerns metropolitan areas. Adding around 10 tC / ha to the 80 tC / ha surface organic layer (humus), the storage capacity of grasslands in the soil is about the same, while the storage capacity in agricultural soil is much less, around 50 tC / ha (Jonard et al. 2017) (Figure4).

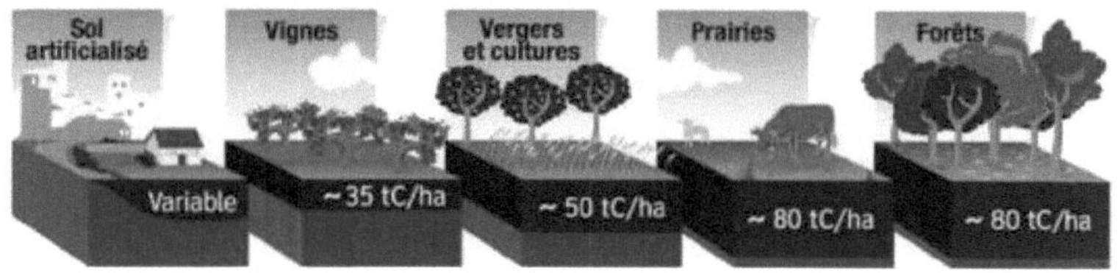

(Source: Martin et al., 2011)

Figure8 : Soil carbon stock over different land uses (0 - 30 cm)

MOS comprises around 55 to 60 percent C by mass. Similar to MOS, COS is divided into several reservoirs based on its physical and chemical stability (FAO and ITPS, 2015; O 'Rourke et al, 2015) :

- **Rapid reservoir** (also labile or active reservoir): After the addition of fresh organic carbon to the soil, the initial biomass is decomposed after 1 to 2 years;
- **Intermediate reservoir**: organic carbon partially stabilized on mineral surfaces and/or protected within aggregates after transformation by microbes. Renewal time is of the order of 10 to 100 years;

- **The slow reservoir** (stable or refractory reservoir): Highly stabilized COS. Very slow renewal (from 100 to over 1000 years);

Carbon storage in soils is temporary, reversible and variable depending on pedological characteristics, organic matter composition, depth, climate humidity, and topography (Chenu et al., 2014).

4.6. The effect of climate change on the COS carbon stock

Temperature and precipitation are the most significant factors in controlling SOC dynamics (Deb et al., 2015). Although an increase in temperature leads to an increase in plant production and therefore increases soil carbon inputs, it also tends to increase microbial decomposition of SOC (Keestrea et al., 2016). A strong empirical conviction conveys the idea that rising temperatures will stimulate the net loss of soil carbon to the atmosphere, leading to a positive terrestrial carbon-climate feedback that could accelerate climate change (Crowther et al., 2016).

Moreover, with climate change, more frequent and heavier precipitation and droughts are to be expected, which could have greater impacts on ecosystem dynamics, compared with the effect alone or combined of rising temperature and CO_2 (IPCC, 2014).

This increase in the frequency of extreme events is likely to exacerbate the quantity and speed of erosion, salinization and other degradation mechanisms, leading to greater carbon losses. Ultimately, climate change, through precipitation, temperature, microorganisms/biotope and vegetation can influence several soil-forming factors, thus affecting the rate of SOC accumulation (FAO and ITPS, 2015).

4.7.   Influence of land use and type of land use on COS

Plant cover has two major benefits for soil carbon:

- They provide and reconstitute humus through the restitution of the biomass produced (aerial and root). The impact on soil organic levels will be beneficial in the short and medium term.
- They optimize soil cover and avoid bare soil. In this way, they boost microbial biomass during the plant's life cycle through rhizodeposition from their roots (Aguer, 2015).

Land use and exploitation generally have a greater impact on soil organic carbon stocks (Aalde et al., 2006). For example, the cultivation of land initially under grassland or forest causes a decrease in stock twice as fast as an increase in stock linked to reverse land conversion.

In Morocco, we have noted that the transformation of forests into cereal-growing areas has led to a loss of over 75% of soil organic carbon concentrations (Zaher et al., 2019). Overgrazing also impacts SOC percentages.

Part 2: Materials and methods

Chapter 1: Introduction to the study area

Introduction

The aim of this work is to highlight the effect of the argan tree reserve (RBA) on the carbon sequestration potential of argan tree ecosystems, particularly in the biogeographical region of the Souss and Dir plains. However, due to time constraints and the vastness of the study area covering more than 730,000 ha, only 2 core zones with their corresponding buffer and transition zones were studied.

1.1.Biogeographical situation of the Arganeraie Biosphere Reserve (ABR)

With a surface area of 2.5 million ha, the RBA is located in south-west Morocco, bordered to the west by the Atlantic Ocean and to the south by Oued Noun and Oued Seyad, while Oued Tensift and Aoulouz mark its limits to the north and east respectively. The study area is characterized by a heterogeneous distribution of argan trees, and extends over a vast and geomorphologically complex physical environment. It contains a very wide diversity of natural environments (altitude, vegetation, soils, climate, surface and underground water, fauna, etc.), giving rise to three major biogeographical units:

- Northern zone: the western High Atlas, from the Argana corridor to the coastline, the southern slopes of the High Atlas from Marrakech to the Essaouira plain;

- The central zone: the Souss plain and Dir ;

- The Southern zone: the Western Anti-Atlas and the Anti Atlas mid-mountain zone;

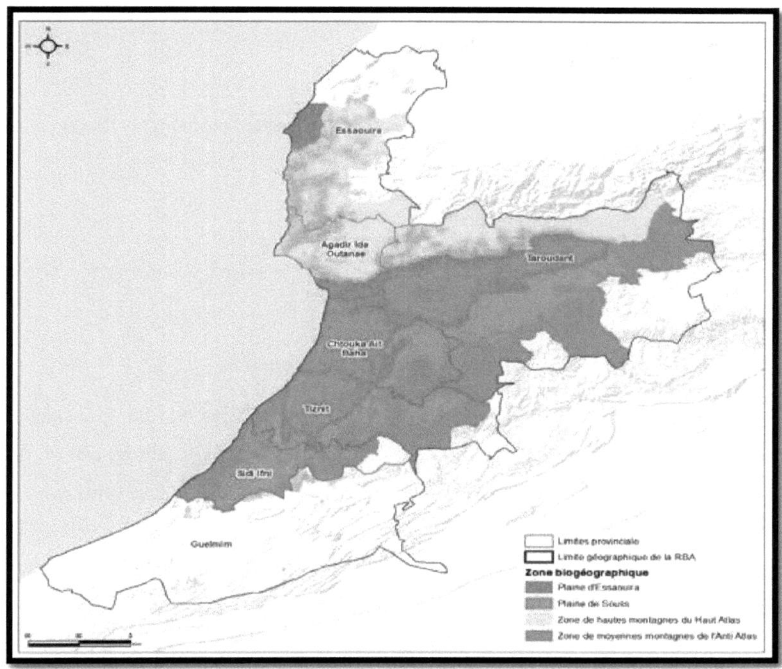

(Source: DEF, 2020)

Figure9 : Map of RBA biogeographical zones

1.2. The choice of study area

As regards the choice of the core zones studied, we based ourselves on the ten-year assessment of the RBA carried out by DEF (2020), which classified the core zones on the basis of the following indicators:

- **Criterion 1: Landscaping**

Landscape value is assessed as a whole, based on an evaluation of the landscape's natural appearance, spectacularity and harmony. The visible presence of modern housing or infrastructure ("modernization") are negative factors, reducing the value of the index assigned.

- **Criterion 2: Plant endemism**

Species with restricted endemism (RBA level) will be valued more highly than species with broader endemism (all of Morocco, Maghreb...).

- **Criterion 3: Animal endemism**

Animal endemism (vertebrates, butterflies) will be assessed according to criteria comparable to those used for flora.

- **Criterion 4: Overall condition of the plant environment**

This criterion is based on the overall state of conservation of the spontaneous vegetation, in particular on the state of the tree cover (problem of felling) and on the state of the pastoral species, a good indicator of the level of grazing (only woody species will be evaluated). Even partial cultivation can only lower the index value. The presence of endangered or rare species, according to available documentation, will increase the value of the index.

- **Criterion 5: Overall condition of fauna**

This criterion is used to assess the overall state of conservation of the fauna, in particular that susceptible to exploitation by man (hunting and poaching). The presence of threatened species (IUCN global and Mediterranean status) will increase the value of the index.

- **Criterion 6: Efforts to regenerate / restore the natural environment**

This human criterion is based on an assessment of various aspects of human activity in relation to the natural environment. It encompasses aspects such as the local population's management of its natural environment, forest planting by the Water and Forestry Service, hunting management, and the existence of a planned or effectively managed Protected Area.

Table1 : Classification of central plain areas

Central zone name	Final score	Comments
PNSM: Mouth of the Oued Massa	**50**	Moderately rich
PNSM: Rwayes Reserve	**43**	Animal reserve
PNSM: Rokkein Reserve	**41**	Animal reserve
Ouameslakht	**31**	Rich
South of Tafingoult	**29**	Rich
Admine	**12**	Although totally degraded, a small non-impacted zone has been delineated (524 ha). Its rating is greatly affected by its surroundings and the many threatening factors.

Although the classification shown in the table above (Table 1) favours the core zones of the SNMP, the inclusion as a core zone of the 2 animal reserves of the Souss-Massa National Park raises questions: a core zone is dedicated to the conservation of the ecosystem and the species that live there. However, these national reserves contain non-native species destined for reintroduction, generally in the Saharan environment.

In a similar way, the central zone of the Oued Massa mouth, whose primary vocation is wildlife conservation, especially bird species, and since our work focuses on quantifying the carbon sequestered in argan tree ecosystems, we have selected the 2 Admine and Ouameslakht zones, which represent the plain and Dir respectively (Figure 9).

The report of the ten-year assessment of the RBA 2008-2017 and the development of the New Action Plan 2018-2027 (DEF,2020) established a description of these 2 zones:

- **The Ouameslakht area**

This site is an argan plantation with a good level of success. The existing argan trees are of good stature, with an average density of 90-100 trees/ha.

The site was clear-cut in the late 1980s and has been protected since then.

- **The Admine zone**

A beautiful lowland forest island, the SIBE part of the Admine forest, although totally degraded, a small non-impacted area has been demarcated (524 ha) with a high density

1.3. Geographical and administrative location of the study area

The administrative situation of the study area is illustrated in the following table:

Table2 : Administrative status of core areas

The central zone	Region	Province	Municipality	Area(ha)
Admine	**Souss-Massa**	Prefecture Inezgane- Ait Melloul	Oulad Dahou	**524**
Ouameslakht		Taroudant	Ida Ougoummad	**72**

Source: DEF (2020)

The study area covers the province of Taroudant and the Inezgane- Ait Melloul prefecture (Figure 10).

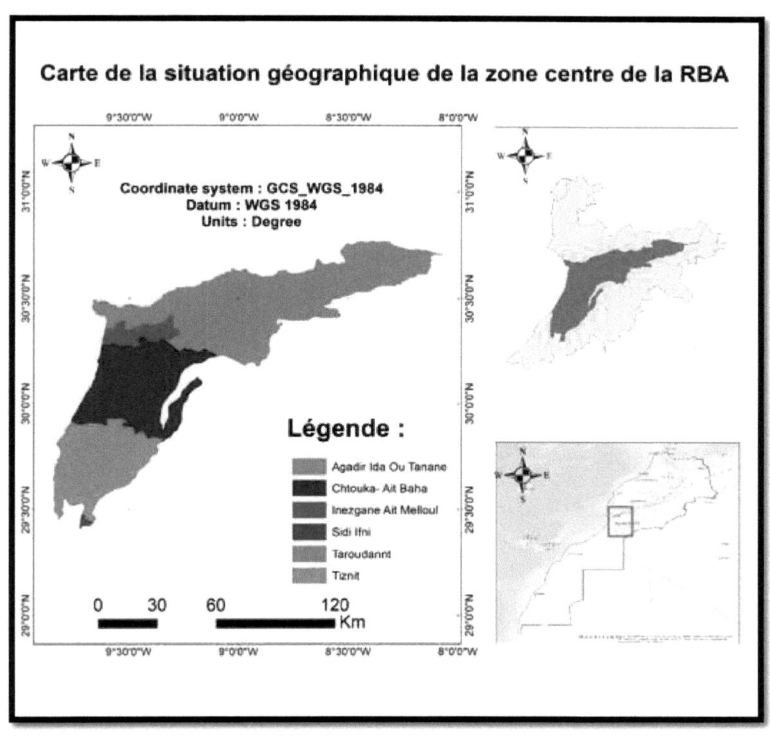

Figure10 : map showing the geographical location of the study area

1.4.Forest situation in the study area

In terms of forestry, the study area is located as follows:

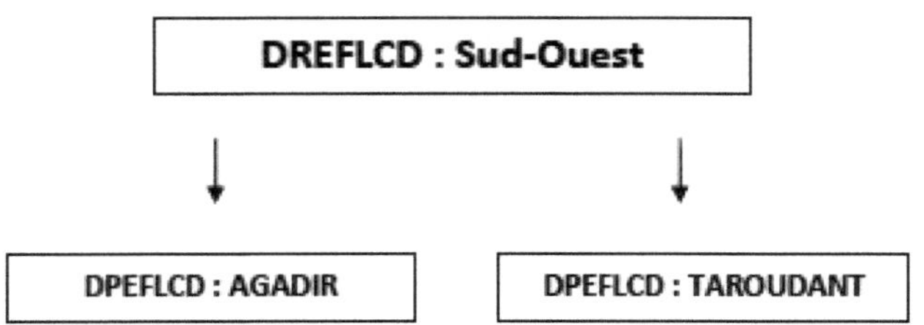

1.5.Physical environment
1.5.1. Geology

The Taroudant plain is relatively flat, with the exception of a few remnants of a Cretaceous Cuesta. It appears to consist of an almost flat expanse that rises from west to east by 700 m over 150 km, with an average gradient of around 5% towards the Anti-Atlas foothills, where the deeply incised alluvial fans offer gradients ranging from 4% at Ouled Taima. Towards the foothills of the Haut-Atlas, the slopes are even steeper (9% at Ouled Taima).

The plain has seen enormous continental accumulations during the Tertiary and Quaternary periods. All the valleys and wadi beds descending from the surrounding slopes deposit their coarse materials at the entrance to the plain, forming the majority of alluvial fans in the mountain foothills.

1.5.2. Soil conditions

The Taroudant plain is marked by open, desert-like landscapes, the result of a sandy soil marked by the presence of hard limestone, which frequently rises to the surface, loosening the scattered clumps of residual vegetation and invading cultivated land, including greenhouse fields, infrastructure and large swathes of the plain's argan grove.

This landscape is also marked by the accumulation of alluvial deposits along valleys and intermittent watercourses (rough mineral soils and less developed soils).

1.5.3. Climate

The RBA is part of the Mediterranean-Saharan transition zone, established around an endemic Moroccan forest species (Argania spinosa), the main feature of the Macaronesian sector, with Mediterranean sclerophyllous forest, woodland and scrub vegetation. According to Emberger's classification, the climate of the area is Mediterranean, arid to semi-arid, with warm and temperate variants.

1.5.4. Precipitation

The spatial distribution of precipitation in the study area is variable and depends essentially on altitude, exposure and distance from the Atlantic coast (Tables 3 and 4). Precipitation reaches :

- 200.3 mm at Admine

- 258.4 mm at Aoulouz

Table3 :Average monthly and annual precipitation at stations in the area

Station	J	F	M	A	M	J	J	A	S	O	N	D	Total
Admine	374	266	360	124	27	03	01	00	13	99	32,1	41,5	200,3
Aoulouz	41,1	38,2	33,8	24,1	9,4	5,1	0,4	1,8	10,8	23,0	30,9	39,8	258,4

Source: (CRF- Rabat, in HCEFLCD, 2006)

Table4 : Average monthly maximum and minimum rainfall

Stations	Max (mm)	Min (mm)	Annual average
Admine	41,5	0,1	200,31
Aoulouz	41 ,1	0,4	258,4

Source: (CRF and Agence du Bassin Hydraulique du Souss Massa, in HCEFLCD, 2006)

1.5.5. Temperatures

Data from the Agadir and Taroudant stations are used, as the study area lies within these 2 provinces. Examination of these data reveals that the hottest months are July and August, while the coldest are December and January.

The average maximum temperature in the hottest month (M) for the Agadir station does not exceed 27°C. The average minimum temperature for the coldest month (m) is around 7.3°C (Table5).

Table5 : Maximum and minimum temperatures at stations in the study area

Station	T(°C)	J	F	M	A	M	J	J	A	S	O	N	D	Moy. Ann.
Agadir	Maxi	20,2	21,5	22,7	23,4	24,1	25	26,4	27	26,7	25,9	23,8	20,9	**24**
	Mini	7,3	8,7	10,7	12,6	14,5	16,5	17,8	18,1	17,2	15,1	12,1	8,6	**13**
Taroudant	Maxi	21,6	23,3	25,5	27,4	29	30,8	35,2	35,8	33,2	29,5	25,7	22,1	**28,3**
	Mini	5,5	7	8,9	10,5	12,2	14,1	16,3	15,9	15,1	13	10	6,4	**11,3**

Source: (CRF- Rabat, in HCEFLCD, 2006)

Thermal amplitude expresses the continental nature of an area and provides information on evapotranspiration. Continentality expresses the distance from the ocean and the conditions associated with it. The greater the thermal amplitude, the greater the continentality. For the stations studied, the thermal range varies from 19.7°C in Agadir to 30.5°C in Taroudant (Table 6):

Table6 : Bioclimatic summaries

Station	T°C Avg. Annual	M	m	M-m	Q2	Emberger climate type
Agadir	18,7	27	7,3	19,7	35	Hot arid
Taroudant	19,8	35,8	5,5	30,5	27	Arid

T: mean annual temperature; M: mean maximum temperature of the warmest month; m: mean minimum temperature of the coldest month; Q2: Amberge rainfall quotient **Source: (CRF- Rabat, in HCEFLCD, 2006)**

1.5.6. The wind

The southwestern Moroccan region is relatively windy, with wind values generally fluctuating between 4 and 6 m/s. The number of days with high winds (>16m/s) averages between 14 and 30 days/year. (Bendaanoun, 1991). Statistics on wind speeds and directions in Agadir show that winds from the west account for 36.7%. These winds are characterized by their coolness, which reduces maximum temperatures by 4°C and increases humidity by 40%. The Chergui (hot, dry wind) accounts for 12.3%, with annual averages of between 30 and 60 days/year. These winds blow from the east, bringing excessive temperatures that can exceed 40°C. WSW and WNW winds account for 10.5% and 9.2% respectively of all wind directions (Bendaanoun, 1991).

1.5.7. Hygrometry

Relative air humidity, when high, is a very important climatic factor for the argan grove, for a multitude of reasons. High air humidity reduces evapotranspiration and, when the dew point is reached, can generate hidden precipitation in the form of dew, which is highly beneficial for vegetation in general and the argan tree in particular. High air humidity, in varying degrees of course, is a key factor in maintaining argan groves (Emberger L. 1925, 1939).

In fact, high air humidity is all the more important as it occurs particularly during the summer period, which considerably reduces evapotranspiration phenomena during this hot, rain-free period, and therefore s The annual averages of relative air humidity as assessed by Bendaanoun (1991) for the southern Moroccan coast and more specifically at Agadir are : 75% (values corresponding to averages established for a long period (1960-1988) and for different observation times during the day (06h + 12h + 18h). Data from internal stations are scarce. Based on estimates by Peltier (1982), the average annual humidity is 64% (06h + 18h).

1.5.8. Drought

In Morocco, 93% of the national territory, under dry climate (arid to subhumid), is characterized by sensitivity to desertification to varying degrees (Neggar,2018). Indeed, the desertification process is increasingly affecting land and remains more pronounced in arid-climate ecosystems, in this case the argan grove, with increasingly long cycles of drought.

This year's drought in Morocco is exceptional in terms of its intensity, extent and duration. The country is experiencing its worst drought in thirty years, with rainfall representing only 13% of the average for the period. The filling of dams rate has never been so low (ONERR, 2022).

The drought has spared neither the argan trees of the plains nor those of the mountains, as the drought of 2022 has affected all the different regions of the country. This unprecedented situation has had an impact on the growth of the herbaceous and shrub layer in argan ecosystems on the plains.

1.6. Natural environment
1.6.1. Floristic biodiversity

The flora of the RBA is characterized by its originality, specificity and diversity. From a phytogeographical point of view, this southern part of Morocco is a crossroads for flora of diverse origins: Mediterranean, tropical, Macaronesian and endemic

The argan tree forms a veritable green bulwark against the Saharan desert and, in addition to its main ecological function, plays an essentially socio-economic role, both for upstream local populations and for the entire value chain that extends beyond its territory, where its ultimate ramifications come to an end

On the plain, where human activity is very intense, the natural vegetation is represented by loose to dense Euphorbia beds dotted with argan trees.

1.6.2. Faunal biodiversity

Specific ecosystem diversity is very high in the RBA area, due to its geographical location between the Atlantic Ocean to the west, the Mediterranean to the north and the Sahara to the south, as well as the altitudinal and climatic variations observed.

The exhaustive fauna inventory was based on a detailed analysis of existing bibliographical data (studies of protected areas, articles, dissertations, work by research...). It revealed a remarkable biodiversity in terms of fauna potential, with a minimum of 66 mammal species, 174 birds, and 63 amphibians and reptiles.

1.7. Human environment

The argan tree on the plains has been the focus of major economic development: urban sprawl, extension of (intensive) crop fields. As a result, the forest area has become a place of conflict, marked by interest and power struggles between players with highly contrasting social representations.

Agricultural investors refer to the Cahier des Charges Générales sur les cultures dans L'arganeraie established in 1983 by the Ministère de l'Agriculture et de la Réforme Agraire. Faced with the development of agricultural investments, the rightful owners, still attached to the values of the argan tree, have created fields fenced off with argan trees and treated as an "agricultural tree" or, more accurately, as a "field tree".

Population growth initially led to the cultivation of more plots of land. And the quest for greater productivity was met from the 50s onwards with the introduction of more efficient ploughing equipment. These included the disc harrow and the brabant harrow, whose use - unlike that of the plough - meant destroying all perennial species. As a result, the number of argan trees remaining on cultivated plots has tended to decline. In a second phase, even greater intensification, this time based on the use of irrigation, was introduced. This followed the strong European demand for out-of-season vegetables at the end of the 60s. The digging of more and more wells gradually led to the colonization of the lowland argan grove by market gardening. The development of cash crops, rather than food crops, such as potatoes (on around a third of the area), watermelons, peppers and above all tomatoes (Benchekroun et al., 1989).

The intensification of farming practices in the lowland arganeraie has also been accompanied by related changes in local production systems. For example, the livestock structure has changed, with sheep, whose numbers have almost doubled over the past 20 years in the Souss region, gradually replacing goats, whose numbers have undergone the opposite trend. The cultivation of alfalfa also attests to the parallel intensification of livestock farming in the area. As for the large nomadic herds, they remain particularly attracted to the argan groves of the plains, which are richer and less densely populated than those of the mountains. Summer camel grazing, which generally lasts from July to September, but can occur earlier in drought conditions, adds to the previous causes of woodland degradation. In addition to being an additional pressure, the damage caused is due to the camels' ability to head off twigs by systematically browsing young shoots. Every year, conflicts arise between local residents and these users.

- **Province of agadir :**

The Agadir area had a population of almost 419,614 in 2004 and 541,118 in 2014 (HCP, 2015), representing a growth rate of 0.029%. The population's way of life is essentially based on livestock and agriculture.

- **Province of Taroudant**

The Recensement général de la population et d'habitat (RGPH) of 2004 attributed a population of 780,661 to the province of Taroudant, representing 25.07% of the region's population. An average density of 47 inhabitants/km^2. The urban environment, essentially the city of Taroudant, is home to 186,471 inhabitants. The rural population totals 594,190, or almost 77% of the province's total population. The majority of rural inhabitants are small-scale farmers and stockbreeders, some of whom engage in small-scale commercial transactions, especially in rural souks.

The area's economy is characterized by the diversity of sectors that form the basis of the population's income resources. Agriculture forms the economic base of the area: modern and traditional, dominated by a high production of citrus fruits, olives, argan oil, vegetables and fodder. It covers a useful area of 175,000 ha, employs a large workforce and attracts substantial investment every year.

Chapter 2. Methodological approach

2.8. Sampling strategy
2.8.1. Sampling method

Since our work is based on the zoning of the ASR at the level of the study area, a stratification of this universe is necessary. This involves dividing the entire site into homogeneous units known as strata, so that an appropriate (random) sampling method can be applied within each stratum. As a result, the sampling units (plots) cover all the environments in the area on an equitable basis. The various steps adopted to carry out this sampling are as follows:

- The two sites (Ouameslakht and Admine) are subdivided into strata or homogeneous units (3 zones per site);

- Set the overall sample size "n";

- Allocate overall size to strata;

- Determining the size of the plot; to do this, we worked on a minimum area, the characteristics of which are developed below.

2.8.2. Determination of strata

In the report on the ten-year assessment of the RBA, the Department of Water and Forests drew up a new zoning scheme to identify homogeneous zones for the 2 sites studied, namely Admine and Ouameslakht.

As a result, we have identified 3 zones for each site, namely the core, buffer and transition zones (Figures 11 and 12).

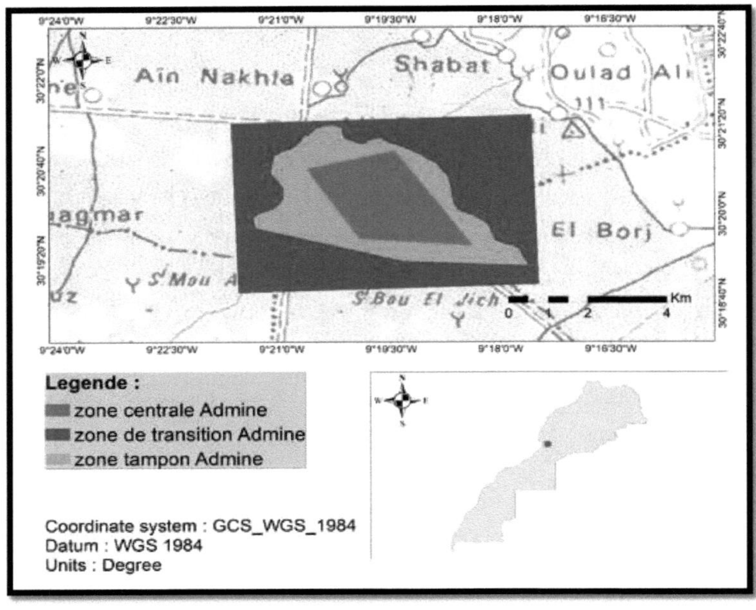

Figure11 : Map of homogeneous units in the Admine study area

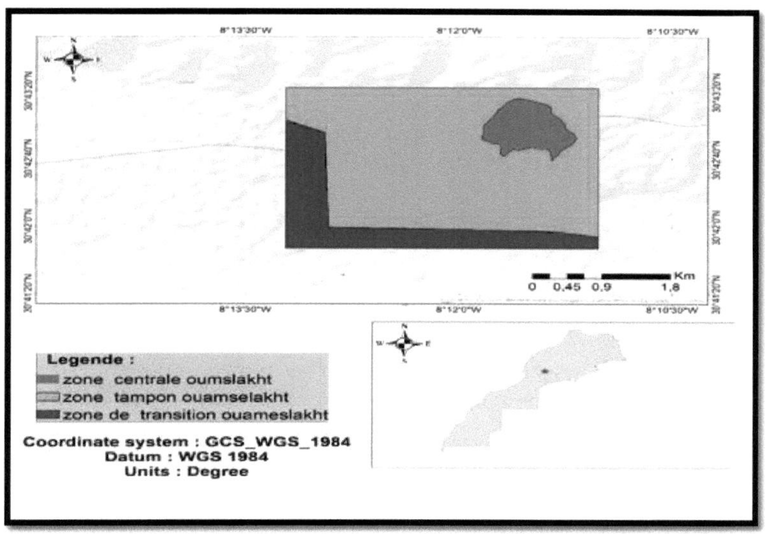

Figure12 : Map of homogeneous units in the Ouameslakht study area

2.8.3. Sample size

Daget and Godron (1982) and Godron (1971, 1976) have shown that 3 to 10 surveys per stratum should be taken into consideration, regardless of stratum size or weight. For this purpose, we selected a sample of 60 plots equally distributed between the strata (Figures 13 and 14).

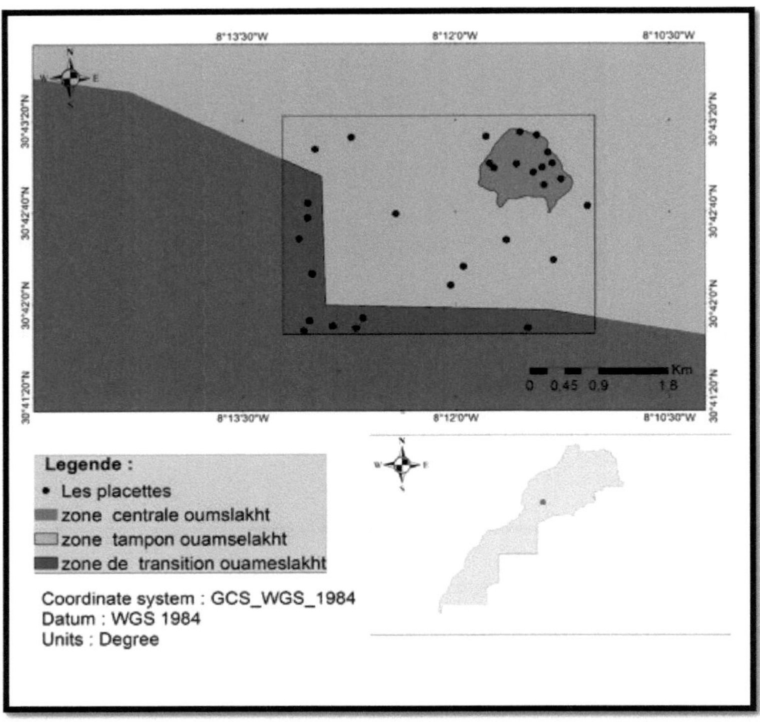

Figure13 : Location map of the plots studied in the Ouameslakht study area

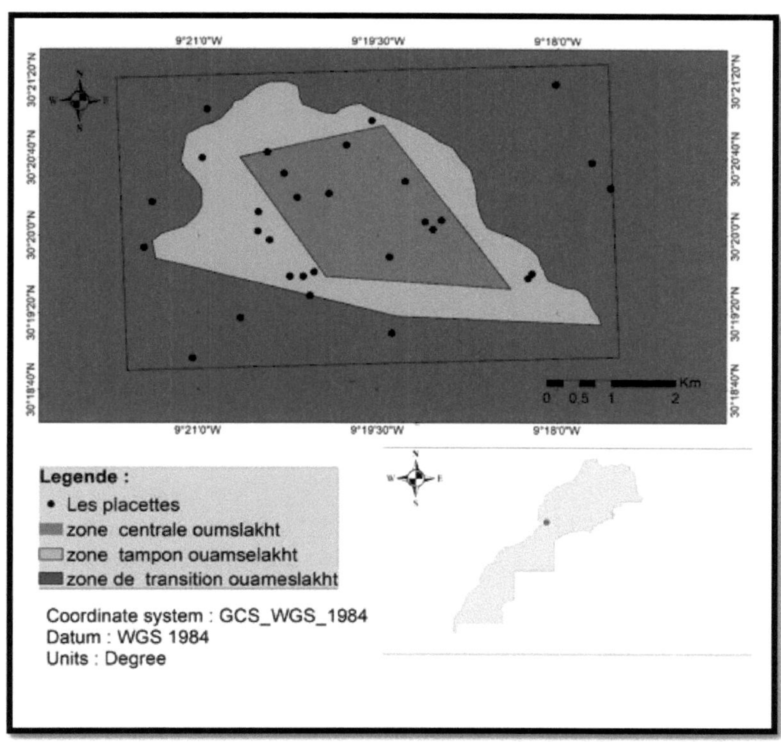

Figure14 : Location map of survey plots in the Admine study area

2.8.4. Sampling plan

In order to obtain measurements that are both accurate and precise, we proceeded by sampling. Our sampling method was inspired by the work of numerous authors (MacDicken, 1997; Hairiah et al., 2001; Woomer et al., 2001; Pearson and Brown, 2005; Mahamane, 2006). Figure 15 illustrates our sampling plan.

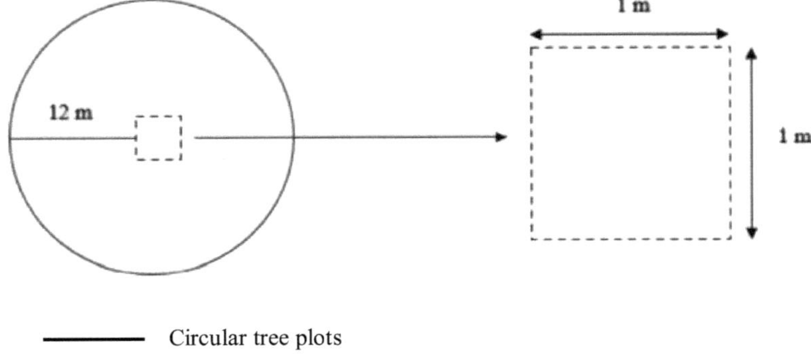

——————— Circular tree plots

- - - - - Square plots for dead wood and 0.25 m^2 for litter

Figure15 : Sampling plan for the different carbon reservoirs

2.9.Methods for estimating C arbone stocks

The choice of methodology will be based on biomass equations. There are two main reasons for this choice.

> The literature review reveals that the use of biomass equations is the most widely used approach for carbon stock estimation;

> Forestry legislation does not authorize the felling of argan trees;

The usefulness of allometry lies in its "non-destructive" approach. Allometric equations enable us to produce quantitative information without destroying the individual.

The mathematical form of allometric equations is mainly a power function:

$$Y= a.X^{\,b}$$

Where:

Y: the characteristic to be measured as a function of X (Biomass or Volume)

X: measured characteristic (circumference, diameter, tree height, etc.)

a: proportionality constant

b: allometric constant

Figure 16 illustrates the general procedure for obtaining the quantity of carbon stored by the argan tree from dendrometric data:

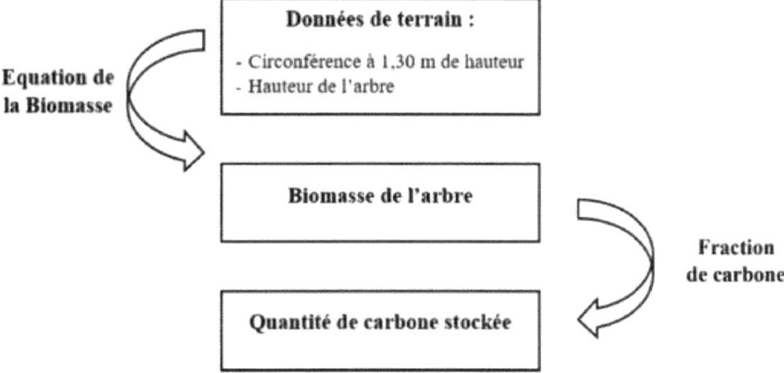

Figure16 : Approach to assessing the amount of carbon stored by the argan tree

2.9.1. Reservoir I: Aerial Biomass (AB)
2.9.1.1. Carbon stock in the tree layer
2.9.1.1.1. Woody biomass
a. Tree sampling

For each stratum, we installed plots of 5 ares each. The parameters sampled within these plots were: circumference at 1.30 m from the ground, total height and average crown diameter.

b. Estimated volume

Given the impossibility of felling argan trees in the forest (an operation not authorized by the local service), we used the timber tariffs recently constructed by the development service of the DREF du Sud-Ouest (2016) and obtained the equation for the 2 zones.

$$V_{wood} = 39.239 - 1.874\,C + 0.037C^2$$

With :

V_{wood}: volume of trunk wood (dm³),

$C_{1.30}$: circumference at 1.30 m (cm),

c. Biomass estimates

The tree's wood biomass is deduced by multiplying the wood volume by its basal density. Tree biomass is calculated as follows:

$$\mathbf{B_A = V_{Wood} \times D}$$

With :

B_A: tree biomass (kg).

V_{Wood}: trunk volume (m^3).

D: specific gravity or infra-density ($kg.m^{-3}$).

- **Determining infradensity :**

Wood density is the ratio between the mass of the wood and its volume. Both vary according to the degree of moisture within the limit of the fiber saturation point (FSP). This parameter is very important, especially as it enables us to calculate the dry mass contained in a volume of wood (Hairiah et al., 2001; Pearson and Brown, 2005). In this work, we used the infradensity, which is simply the quantity of dry matter contained in a volume of wood.

Wood samples collected in the field were placed in boiling water and weighed regularly until their weights stabilized.

At this stage, the sample has reached the fiber saturation point, which corresponds to its saturated weight. The samples were then placed in an oven and weighed regularly until a stable dry weight was reached. The expression for infradensity is as follows:

$$\mathbf{Db = \frac{1}{\frac{ms}{mo} - 0{,}347}}$$

With :

Db: infradensity or basal density ($g\ cm^{-3}$),

ms: mass of sample in saturated state,

50

mo: mass of sample after baking.

0.347: constant

After determining the mass of the samples in the saturated state and after drying, we estimated the infra-density of the wood from the plain argan tree at **930 kg/m³**.

> d. Estimation of carbon stock in above-ground biomass

Converting biomass into carbon requires a conversion factor. This is the fraction of carbon in biomass, generally equal to 0.5 (MacDicken, 1997; Woomer et al., 2001; Pearson and Brown, 2005). However, this fraction varies from species to species. According to belghazi. T et al. (2017), the percentage by weight of carbon contained in dry matter for argan trees is **48.5%.**

$$\textbf{StC}_{\textbf{BA}} = \textbf{BA x 48.5}$$

With :

StC$_{BA}$: carbon stock in kg in the above-ground biomass of an individual tree,

BA: above-ground biomass in kg.

2.9.1.1.2. Leaf biomass

With regard to leaf biomass for argan trees, we selected the allometric model constructed by Aamou (2013), which takes the form:

$$\textbf{PSF} = \textbf{0.53 Dm}^{1.361}$$

PSF: Dry leaf weight (Kg)

Dm: Average diameter of crown to ground projection (m)

Using the calcination method, we found that the weight percentage of carbon contained in the leaf biomass is **48.5%.**

2.9.2. Reservoir 2: Biomasse Souterraine (BS)

Root carbon stock, considered an important biomass of carbon, is very difficult to determine. Based on data reported in the literature, it can be estimated at between 15 and 24% of above-ground carbon stock (Arrouays et al., 1999). Belghazi. T et al.

(2017) adopted an average percentage of 20% of aerial carbon stock for argan stands in the haha plateau.

Given that Belghazi T et al (2017) have worked on the argan tree, so for our case study, we adopted an average percentage of 20% of the aerial carbon stock .

2.9.3. Tank 3: Necromass (N)
2.9.3.1. Dead wood
a. Sampling dead wood

- **Dead wood lying**

In the circular plot, a 1m x 1m quadrat was placed near the argan tree to collect dead wood. Typical deadwood samples were collected to determine the dry mass of deadwood lying on the ground, and to assess the basal density of deadwood.

b. Carbon stock estimation

Carbon stock in dead wood is calculated as follows :

$$StC_{BM} = MS_{BM} \times 48.5$$

With :

StC_{BM}: carbon stock in dead wood ;

$MS_{(BM)}$: dry mass of dead wood in Kg

2.9.3.2. Litter
a. Litter sampling

Within the 0.25 m² plot, all litter was collected and returned to the laboratory to determine their dry mass.

b. Biomass estimates

The litter samples were oven-dried until the dry weight had stabilized.

c. Carbon stock estimation

Carbon stock in litter is calculated as follows:

$$StC_{L} = MS_{L} \times 48.5$$

With :

StC$_L$: carbon stock in litter (Kg) ;

MS$_L$: dry mass of litter (Kg).

2.9.4. Tank 4: soil (S)
2.9.4.1.Soil sampling :

Soil samples were taken to determine bulk density. Samples were taken using 200 cm metal cylinders[3]. Three replicates per stratum were selected for all samples.

Other soil samples were taken to determine the quantity of organic carbon stored in the soil, and we limited ourselves to 30 cm depth, because it is rich in organic matter, micro-organisms and root concentration. Most of the calculations made in previous work refer to this horizon, which is also the reference horizon for IPCC work (Pellerin et al., 2019)

Harvested soil samples, like all other samples, were immediately stored in sealed plastic bags.

2.9.4.2.Physico-chemical soil characterization
a. Apparent density

Soil bulk density is defined as the ratio between the mass of dry soil (usually oven-dried at 105°C) and its fresh volume. It reflects the soil's ability to function for structural support, water and solute movement and soil aeration. It is determined by collecting soil cores using a cylinder of known volume (200 cm^3). The fresh soil cores are oven-dried at a temperature of 105°C for a period of 24 hours, after which their dry weight is determined. The volume of fresh soil is also calculated by determining the volume of the cylinder used. Bulk density is calculated using the following formula:

$$\mathbf{Da} = \frac{\mathbf{Ms\ du\ sol\ sec}}{\mathbf{VT\ du\ sol\ frais}}$$

With

Da: Bulk density (g cm^{-3})

Ms of dry soil: mass of dry soil (g)

V_T of soil: Total volume of soil (g cm^{-3})

b. Total organic carbon

Quantification of organic C in soil samples is performed using the **Walkley-Black (1934)** chromic acid wet oxidation method. Oxidizable soil organic carbon is oxidized by a solution of potassium dichromate in concentrated sulfuric acid. The heat of the reaction increases the temperature, which is sufficient to induce substantial oxidation.

The chemical reaction is as follows:

$$2\ K_2Cr_2O_7^{2-} + 3\ CO + 16\ H^+ \rightarrow 4\ Cr^{3+} + 3\ CO_2 + 8\ H_2O$$

The Cr_2O7^{2-} reduced during the reaction with soil is proportional to the oxidizable organic C present in the sample. Organic carbon can then be estimated by measuring the remaining unreduced dichromate by back titration with ferrous sulfate, using the o-phenanthroline-ferrous complex as an indicator.

$$6\ Fe^{2+} + Cr_2O_7^{2-} + 14\ H^+ \rightarrow 2\ Cr^{3+} + 6\ Fe^{3+} + 7\ H_2O$$

> **Reagents**

- Potassium dichromate $k_2Cr_2O_7$ 1N
- Iron sulfate ($FeSO_4$ .7H) 0.5N
- Concentrated sulfuric acid 96% (H2SO4) (density =1.83)
- O-phenanthroline-ferrous complex 0.025M

> **How it works**

After drying at 65°C for 48 hours, the samples were crushed and sieved to 2 mm. Soil samples were processed in the following steps to deduce the amount of organic carbon sequestered in the soil:

- ✓ Weigh out 1g of soil sieved to 2 mm or 0.5g if the soil is rich in organic matter.
- ✓ Place the soil sample in a 250 ml Erlenmeyer flask.
- ✓ Add 10 ml 1N potassium dichromate and shake by hand until the soil has diffused.

- ✓ Add 20 ml concentrated sulfuric acid, stirring until the chemicals are mixed for 1 min.
- ✓ Leave to stand for 30 min under a high heat lamp.
- ✓ Add 200 ml distilled water, homogenize well and leave to stand for at least 2 hours.
- ✓ Then transfer 50ml of the solution into a 250ml beaker.
- ✓ Add 3 to 4 drops of ferrous O-phenanthroline indicator; the solution takes on a greenish to dark green hue.
- ✓ Titrate excess dichromate with iron sulfate (FeSO4) drop by drop until a red-brown color appears (Figure 19 and 20).

The total organic carbon contained in the soil is calculated using the following formula:

$$C\,(\%) = 5,85 * \frac{(Vt - Vi)}{p * Vt}$$

Where:

C: soil organic carbon concentration (%) ;

Vt: volume of Mohr's salt solution corresponding to the titration of the control (ml) ;

Vi: volume of Mohr's salt solution corresponding to sample titration (ml) ;

P: weight of soil sample taken (g) ;

- **Estimation of soil carbon stock**

Carbon percentage and bulk density are used to determine carbon and soil organic matter stocks. Carbon stock in the 4th reservoir is calculated using the following formula:

$$StC_S = (Da \times E \times Corg)*100$$

StC$_S$: soil carbon stock (t C. ha^{-1}) ;

Corg: concentration of organic C in the horizon (%) ;

Da: soil bulk density (g. cm^{-3}) ;

E: effective horizon thickness (cm), excluding stones with a diameter greater than 6 cm.

2.10. Statistical analysis

Calculations of dry mass and carbon stocks in carbon reservoirs and graphical representations were carried out using Excel software (version 2016).

All statistical analyses were performed using SPSS 25 software. The means of the different variables were compared using the ANOVA1 single-criterion analysis of variance method. Tukey's honestly significant difference (HSD) procedure was used for pairwise comparisons. Differences are assessed at the 5% probability level.

2.11. Diachronic study of land use evolution

To study the evolution of land use in the RBA and in the Souss et Dir plain biogeographical zone over a 23-year period (between 1998 and 2021), in order to create a Land Use Change Map, we adopted a methodology based on unsupervised classification using the "Trends.Earth" tool on Qgis software version (3.16).

As for the evolution of land use in the two central zones, we adopted supervised classification using the Google Earth Engine platform, which enabled us to analyze Landsat satellite images from 1998 and 2021.

2.11.1. Data used

Our satellite image dataset consists of 2 Landsat images from 1998 and 2021. The choice of images was based essentially on their dates of acquisition in the year and the percentage of clouds, so as to have images acquired in close proximity to each other and a low percentage of clouds. We opted for 1998, the date on which the RBA was created, and 2021, the most recent date on which satellite images are available (Table 7).

Table7 :Characteristics of Landsat satellite images

Satellite	Pick-up date	Spatial resolution (m)
Landsat 5	1998	30
Sentinel	2021	10

2.11.2. Image classification

The chosen classification method is supervised classification. Based on our knowledge of the terrain and the spectral signature, we defined the class to which each pixel in our image belongs. The algorithm used is maximum likelihood, based on Bayes' rule, which calculates for each pixel its probability of belonging to one class rather than another. The pixel is assigned to the class with the highest probability of belonging.

In this study, five land-use classes were defined (Table 8):

Table8 : Land use classes

Class	**Description**
Forests	This class essentially comprises natural forests, artificial forests (reforestation) and matorrals (shrubland).
Farming	This class includes irrigated crops (medium and large fields of irrigated cropland and grass), rainfed crops (bour) and agroforestry (permanent crops, agricultural plantations).
Built environment	The class includes urban areas, infrastructures and related developed areas.
Bare floors	Bare land without vegetation (barren), rocks, dunes...
Course	These include natural pastures and meadows, sparsely vegetated areas, and natural vegetation associations and mosaics.
Water	This category includes peat extraction areas and land that is covered or saturated with water for all or part of the year and does not fall into the

	categories of forest land, cropland, grassland or settlements. It includes reservoirs as an exploited subdivision and natural lakes and rivers as non-exploited subdivisions.

2.11.3. Change detection and spatialization

After supervised classification of the two images studied, we obtained the surface areas of each class, and then made a comparison between the classes to quantify the evolution of land use between 1998 and 2021.

Part 3: Results and discussion

Chapter 1: Carbon stocks in carbon reservoirs

Introduction

In this chapter, we discuss the carbon stock in above-ground biomass, below-ground biomass and necromass according to RBA zoning. Above-ground biomass comprises only tree biomass, due to the absence of shrub and herbaceous strata during February, as a result of the unprecedented drought that Morocco is experiencing in 2022, while below-ground biomass consists of root biomass. Necromass comprises the carbon stock contained in both dead wood and litter.

1.1. Carbon stock in above-ground biomass

The calculation of woody and foliar biomass at the two sites studied is shown in the following table (Table 9):

Table9 : Average tree biomass by zoning, 2 sites

Above-ground tree biomass	Average biomass (t DM. ha^{-1})					
	Admine			Ouameslakht		
	Central	Tampon	Transition	Central	Tampon	Transition
Woody	12,00	8,19	8,98	8,48	4,93	2,92
Foliar	0,481	0,32	0,36	0,27	0,16	0,10
Total	12,48	8,51	9,34	8,75	5,10	3,02

The stands studied are characterized by a stump-trunk system. Analysis of the table shows that the total biomass of the stands studied varies from 3.02 to 12.48 tonnes of dry matter per hectare (DM/ha), with average density ranging from 90 to 100 trees/ha. However, woody biomass ranges from 2.92 to 12 tonnes DM/ha.

There is little literature on the dry biomass of the argan tree. However, Belghazi. T et al (2016) found average biomasses ranging from 1.62 to 10.2 tonnes of dry matter per hectare (DM/ha) for an average density ranging from 75 to 118 trees/ha in a stand of young argan coppice in Haha.

The results show that the average biomass per hectare is high in the central zones for both sites. At Ouameslakht, the above-ground biomass of the argan tree varies from 3.02 tonnes MS/ha in the transition zone to 8.75 tonnes MS/ha in the central zone, while at Admine, the average biomass varies from 8.51 tonnes MS/ha in the buffer zone to 12.48 tonnes MS/ha in the central zone

At Admine, woody biomass in the core, buffer and transition zones represents 96.15%, 96.24% and 96.15% of above-ground biomass respectively.

Similarly for Ouameslakht, woody biomass in the core, buffer and transition zones represents 96.91%, 96.67% and 96.69% of above-ground biomass respectively (Figure 17).

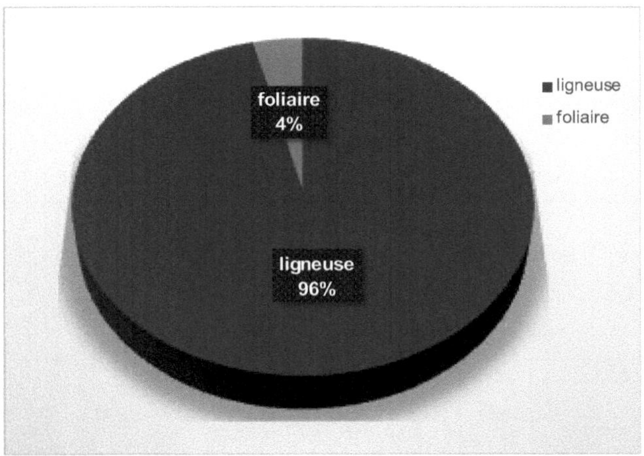

Figure17 : Contribution of leaf and woody biomass to total above-ground biomass

There is little literature on the foliar biomass of the argan tree. In Table 10, we have grouped together those relating to work carried out on the argan tree. This comparison shows that the argan tree stands at the 2 sites on the Souss plain have a significant leaf production potential, with leaf production ranging from 0.10 to 0.48 tonnes of dry matter.

Table10 : Comparison of argan leaf biomass production

Author	Website	Species	Age	Density (Spruce/ha)	Leaf biomass in t MS / ha
Benzyane (1989)	Plateau Haha	Argan tree	Aged stand	295	1,65
Oudaha (2007)	Plateau Haha	Argan tree	Young rejects	51	0,12
Marouch (2013)	North face of Jbel Amssiten	Forest on Argan tree stump	Aged stand	100	1,35
Aamou (2013)	South face of Jbel Amssiten	Forest on Argan tree stump	Aged stand	100	1,36
OUHA (2022)	Admine	Forest on Argan tree stump	Aged stand	100	0,35

The carbon stock in the tree biomass of argan forests at the 2 sites was estimated according to the RBA zoning and the results are presented in Table 11 and Figure 18.

Table11 : Aerial carbon stock in the tree stratum by zoning at the 2 sites

Website	Zone	Dry mass (t/ha)	Carbon stock (t/ha)
Admine	Central	12,48	6,05

	Tampon	8,51	4,13
	Transition	9,34	4,53
Ouameslakht	Central	8,75	4,24
	Tampon	5,10	2,47
	Transition	3,02	1,46

At the Admine site, the carbon stock in tree biomass varies from 4.13 to 6.05 (t/ha). At Ouamslakht, the carbon stock in tree biomass varies from 1.46 to 4.24 (t/ha). This difference is linked to the volume of trees per hectare.

The lack of data on carbon sequestration by argan ecosystems means that only limited comparisons can be made. Nevertheless, these results are consistent with those reported in the literature. Indeed, the only data available on argan trees are those of Belghazi and Ouswati (2016), who estimated the carbon sequestered in a stand of argan coppice in the Haha plateau at 5 tonnes C/ha.

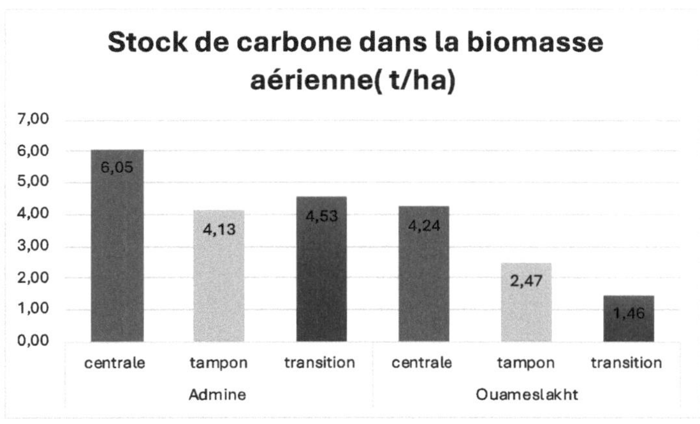

Figure18 : Aerial carbon stock in the tree stratum by zoning at the 2 sites

Analysis of variance with a single classification criterion showed a significant difference between the means of carbon content according to the different zones of the

Ouameslakht site (Table 12). We can say that there is a significant effect of zoning on the carbon content stored in the different Ouameslakht zones ($p < 0.05$).

In fact, the post hoc multiple comparison of means test for carbon stocks at the Ouameslakht site revealed significant differences between the central zone and the transition zone, while the difference in means shows the superiority of the central zone over the other two zones (Table13).

This is generally due to the fact that the central zone is fenced off, allowing the above-ground biomass of the argan tree to develop.

Table12 : ANOVA 1: tree stratum at Ouameslakht site

Carbon (t / ha)					
	Sum of squares	ddl	Medium square	F	Sig.
Factor	0,091	2	0,046	4,413	**0,022**
Residue	0,280	27	0,010		
Total	0,371	29			

Table 13: Multiple comparisons (two by two) of average carbon content values

	(I) the	(J) the	Average difference (I-J)	Standard error	Sig.	95% confidence interval	
						Lower terminal	Upper terminal
Tukey's significant difference	Central	Tampon	0,088	0,045	0,145	-0,024	0,201
		Transition	0,132[*]	0,045	0,019	0,019	0,245
	Tampon	Central	-0,088	0,045	0,145	-0,201	0,024
		Transition	0,044	0,045	0,601	-0,068	0,157

| | | Central | -0,132* | 0,045 | 0,019 | -0,245 | -0,019 |
| | Transitio n | Tampon | -0,044 | 0,045 | 0,601 | -0,157 | 0,068 |

*. The mean difference is significant at the 0.05 level.

As for the Admine zone, the analysis of variance with a single classification criterion showed that there was no difference between the means of carbon content in the different zones of the site (Table 13.2). This could be the result of anthropogenic action amplified by new forms of land use that are altering the carbon stock and stock renewal potential in the central zone of this site.

Table13 : ANOVA 1: tree stratum at the Admine site

Carbon (t C /ha)					
	Sum of squares	ddl	Medium square	F	Sig.
Factor	0,093	2	0,047	2,716	**0,084**
Residue	0,463	27	0,017		
Total	0,557	29			

1.2. Carbon stock in underground biomass

With regard to the carbon stock in the underground biomass, a percentage of 20% of the above-ground carbon stock has been adopted as explained in previous sections (see section .., paragraph ,page 45 of this document).

The carbon stock contained in the underground biomass of the argan tree was determined for each site, and the results are presented in Table 14.

Table14 : Amount of carbon stored in roots

Website	Zone	Underground biomass carbon stock (t C/ ha)
Admine	Central	1,21
	Tampon	0,83

	Transition	0,91
Ouameslakht	Central	0,85
	Tampon	0,49
	Transition	0,29

Since the carbon stock in the roots is correlated with that of the above-ground biomass, the analysis of the results is similar to that of the above-ground biomass.

1.3. Carbon stock in necromass

1.3.1. Carbon stock in dead wood

No standing dead wood was found at any of the sites. The carbon stocks presented here (Table 15 and Figure 19) are exclusively those lying on the ground.

Table 15 shows the results of carbon sequestration in dead wood.

Table15 : Quantity of carbon stored in dead wood

Website	Zone	Carbon stock in dead wood (t C/ ha)
Admine	**Central**	0,51
	Tampon	0,27
	Transition	0,47
Ouameslakht	**Central**	0,41
	Tampon	0,21
	Transition	0,27

Table16 and figure19 show that the central zones of Admine and Ouameslkaht sequester the most carbon in dead wood, and according to Derrière et al. (2012), the stock of dead wood, and therefore its carbon, increases as the aerial carbon stock of living trees increases.

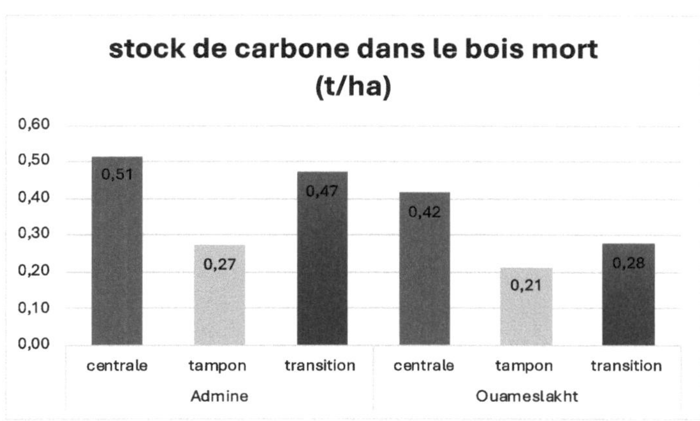

Figure19 : Carbon stock in dead wood by zoning at the 2 sites

However, analysis of variance ANOVA 1 shows that there is no significant difference between the zones at the two sites (p> 0.05) (Tables 16 and 17). This could be explained by the collection of dead wood by the riparian population and the low input of dead wood at both sites.

Table16 : ANOVA 1 : dead wood in Admine site

Carbon stock in dead wood					
	Sum of squares	ddl	Medium square	F	Sig.
Factor	0,001	2	0,001	0,528	**0,615**
Residue	0,006	6	0,001		
Total	,0007	8			

Table17 :ANOVA 1: dead wood in Ouameslakht site

Carbon stock in dead wood					
	Sum of squares	ddl	Medium square	F	Sig.
Factor	0,001	2	0,000	1,674	**0,264**
Residue	0,001	6	0,000		

Total	0,002	8			

1.3.2. Carbon stock in litter

Determining the carbon stock in litter is a very important parameter, as litter represents one of the main sources of soil carbon. The return of organic matter in the form of litter is very important in soil carbon enrichment, as these two parameters are correlated. Litter is therefore one of the main sources of soil carbon (Quenea, 2004).

We therefore estimated the carbon stock in the different zones for both sites and the results are presented in the following table 18 and figure 20:

Table18 : Amount of carbon stored in litter

Website	Zone	Carbon stock in litter (t / ha)
Admine	Central	0,11
	Tampon	0,29
	Transition	0,16
Ouameslakht	Central	0,46
	Tampon	0,43
	Transition	0,18

The distribution of carbon stock in litter is as follows: the central zone of Ouameslakht sequesters the most carbon with around 0.46 t/ha, followed by the buffer zone with 0.43 t/ha (Figure 20).

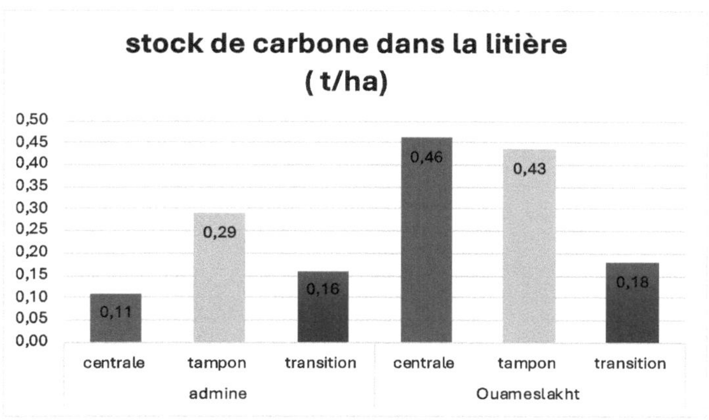

Figure20 :Carbon stock in litter according to zoning in the 2 sites

ANOVA1 results (Table 19) show that in the Admine site, zoning has no significant effect on litter carbon stock (p> 0.05). In this respect, overgrazing could be responsible in the rangelands and cultivation in the central zone.

Table19 : ANOVA 1 : ANOVA 1 test results for litter in different areas of the Admine site

Carbon stock in litter					
	Sum of squares	ddl	Medium square	F	Sig.
Factor	0,052	2	0,026	1,120	**0,386**
Residue	0,140	6	0,023		
Total	0,192	8			

However, at the Ouameslakht site, ANOVA 1 analysis shows that zoning has a significant effect on litter carbon stock (p < 0.05)

The multiple comparison test (Table 20) shows that at this site, the carbon content in the litter in the central zone is not significantly different from that in the buffer zone. However, the carbon content in the core zone is significantly different from that in the transition zone.

The isolated nature of the central zone and the absence of anthropogenic activities would have greatly influenced the high carbon stock values of the zone compared to those of other zones.

Table20 : ANOVA 1 : litter in Ouameslakht site

Carbon stock in litter					
	Sum of squares	ddl	Medium square	F	Sig.
Factor	,145	2	,073	5,675	**0,041**
Residue	,077	6	,013		
Total	,222	8			

Table 20: Multiple comparisons (two by two) of average carbon content values

Dependent variable: carbon stock in litter (t/ha

LSD

(I) the	(J) the	Average difference (I-J)	Standard error	Sig.	95% confidence interval	
					Lower terminal	Upper terminal
Central	Tampon	0,025	0,092	0,790	-0,200	0,251
	Transition	0,281[*]	0,092	0,023	0,055	0,507
Tampon	Central	-0,025	0,092	0,790	-0,251	0,200
	Transition	0,255[*]	0,092	0,032	0,029	0,481
Transition	Central	-0,281[*]	0,092	0,023	-0,507	-0,055
	Tampon	-0,255[*]	0,092	0,032	-0,481	-0,029
*. The mean difference is significant at the 0.05 level.						

1.4. Soil carbon stock

From a global perspective, soils are an important terrestrial carbon reservoir. Soil organic carbon (SOC) is the main constituent of soil organic matter, which in turn is an important determinant of soil fertility, water storage capacity and biological activity. It also influences soil compactibility, friability and aggregation, which are directly linked to soil permeability and erodibility. Estimating the amount of soil carbon is therefore of great interest.

Table 21 shows the results for the quantities of carbon sequestered in the soil.

Table21 : Quantity of carbon stored in the soil

Website	Zone	Soil carbon stock (t/ ha)
Admine	**Central**	57,12
	Tampon	45,21
	Transition	13,93
Ouameslakht	**Central**	38,79
	Tampon	28,24
	Transition	17,46

According to Table 21 and Figure 21, the central zones of the 2 Admine and Ouameslakht sites sequester more carbon, at 75.94 t C/ha and 38.79 t C/ha respectively.

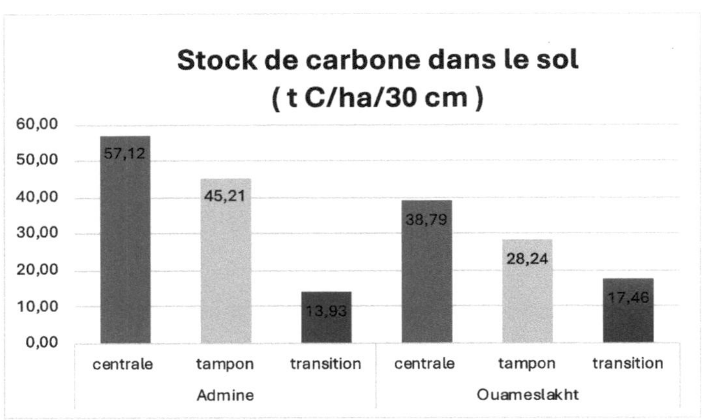

Figure21 : Soil carbon stock by zoning at the 2 sites

The results presented in the graphs show that the amount of carbon sequestered in the soil over the first 30 cm is greater in the central zones. However, ANOVA1 results (Table 22) show that soil carbon stock is not significantly different between zones at the Admine site (p >0.05).

Table22 : ANOVA 1 : soil in Admine site

Carbon stock t/ha					
	Sum of squares	ddl	Medium square	F	Sig.
Factor	1523,706	2	761,853	,902	**0,454**
Residue	5067,168	6	844,528		
Total	6590,874	8			

As for the Ouameslakht zone, analysis of variance shows that there is an effect of zoning on soil carbon sequestration (Table 23), so a multiple comparison of means is required.

Table23 : ANOVA 1: soil at Ouameslakht site

Carbon stock t/ha					
	Sum of squares	ddl	Medium square	F	Sig.

Factor	546,728	2	273,364	8,524	**0,018**
Residue	192,415	6	32,069		
Total	739,143	8			

The multiple comparison of means revealed significant differences between the central zone and the transition zone. The difference in means shows the superiority of the central zone over the other two zones (Table 24).

The practices implemented could modify the carbon stock by altering the type of land use, for example by cultivating forested land and reducing soil cover. However, fencing has ensured the return of some plant biomass to the soil by reducing the intensity of grazing and limiting tillage operations that stimulate the decomposition of organic matter.

According to current knowledge (Pellerin et al, 2019), among these practices, it is those that influence the quantities of carbon returned, which feed the soil's organic carbon pool, that will determine the equilibrium carbon stock in a given pedoclimatic context.

Table24 : Multiple comparisons (two by two) of average carbon content values

Dependent variable: soil carbon stock (t C/ha							
	The zone (I)	The zone (J)	Average difference (I-J)	Standard error	Sig.	95% confidence interval	
						Lower terminal	Upper terminal
LSD	Central	Tampon	11,080	4,623	0,054	-0,233	22,394
		Transition	19,004[*]	4,623	0,006	7,690	30,318
	Tampon	Central	-11,080	4,623	0,054	-22,394	0,233
		Transition	7,924	4,623	0,137	-3,389	19,23
	Transition	Central	-19,004[*]	4,623	0,006	-30,318	-7,690

| | | Tampon | -7,924 | 4,623 | 0,137 | -19,238 | 3,389 |

*. The mean difference is significant at the 0.05 level.

1.5. Total carbon stock and its distribution in reservoirs

After calculating carbon stocks in the various reservoirs (above-ground biomass, below-ground biomass, necromass and soil), the total stock in the lowland argan tree ecosystem is given in the following table:

Table25 : Total carbon stock in reservoirs

Sites	Zone	Tree stock (T/ha)	Root stock (T/ha)	Deadwood stock (T/ha)	Litter stock (T/ha)	Soil stock (T/ha)	Stock Total (T/ha)
Admine	Central	6,05	1,21	0,51	0,11	57,12	65
	Tampon	4,13	0,83	0,27	0,29	45,21	50,73
	Transition	4,53	0,91	0,47	0,16	25,56	31,63
Ouameslakht	Central	4,24	0,85	0,41	0,46	38,79	44,75
	Tampon	2,47	0,49	0,21	0,43	28,24	31,84
	Transition	1,46	0,29	0,27	0,18	17,46	19,66

In the Admine site, the total carbon stock of the different reservoirs is 49.12 t C ha^{-1} and 32 t C ha^{-1} for the Ouameslakht site. The total carbon stock contained in each zone is as follows (Figure 22):

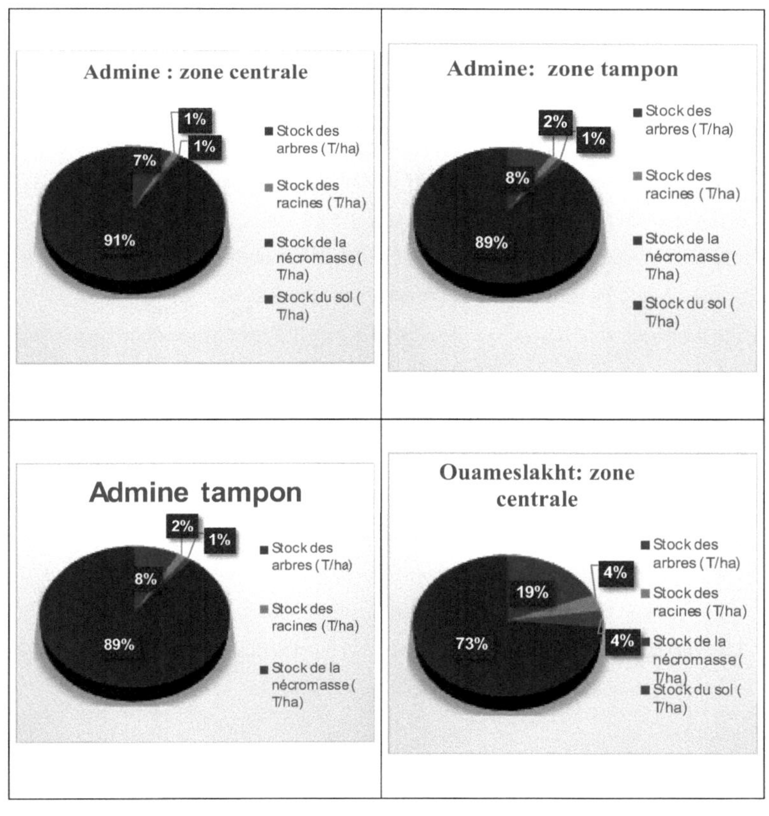

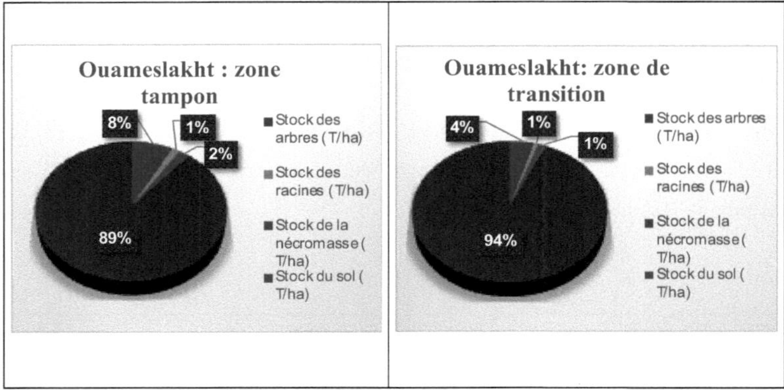

Figure22 : Carbon distribution in the various reservoirs

- On the Admine site:

✓ The central zone stores 65 t C/ha, of which 9% in above-ground biomass, 2% in below-ground biomass, 1% in necromass and 88% in soil.

✓ The buffer zone stores 50.73 t C/ha, of which 8% in above-ground biomass, 2% in below-ground biomass, 1% in necromass and 89% in soil.

✓ The transition zone stores 31.63 t C/ha, of which 8% in above-ground biomass, 2% in below-ground biomass, 1% in necromass and 89% in soil.

- At the Ouameslakht site:

✓ The central zone stores 44.75 t C/ha, of which 19% in above-ground biomass, 4% in below-ground biomass, 4% in necromass and 73% in soil.

✓ The buffer zone stores 31.84 t C/ha, of which 8% in above-ground biomass, 1% in below-ground biomass, 2% in necromass and 89% in soil.

✓ The transition zone stores 19.66 t C/ha, of which 4% in above-ground biomass, 1% in below-ground biomass, 1% in necromass and 94% in soil.

In general, soil is a key reservoir for carbon and can hold up to two-thirds of total ecosystem carbon stocks (Dixon et al., 1994; Mao et al., 2010; IPCC, 2014). This has also been demonstrated in Mediterranean forest ecosystems (Ruiz Peinado et al., 2013). In our context, the top 30 centimetres of soil, represent the main carbon storage compartment of this ecosystem with around 85.1% of total average organic carbon stocks. The above-ground biomass of the argan tree is the second largest carbon reservoir with only 10.6%, the below-ground biomass comes third with 2.5% and finally the necromass whose contribution to the total organic carbon stock represents 1.6%.

Carbon distribution at reservoir level apparently depends on the ecosystem. Indeed, it has been reported that tropical forests have 56% of the carbon stock in biomass and 32% in soil, while boreal forests have only 20% in biomass and 60% in soil; the remaining carbon is retained in dead wood and litter (Pan et al., 2011).

By comparison with other Mediterranean ecosystems, the average total carbon stock in our study area is low compared with that of cork and cedar forests, where it can reach 140 and 300 tonnes C/ ha respectively (Rhoufacha, 2021; Hounzandji, 2008). This variability could be explained by climate, stand structure (density, age, etc.) and soil type.

1.6.   Converting carbon into carbon dioxide

Carbon stock has been described in terms of biomass and carbon. When we talk about climate change, we're interested in emissions from land-use change in terms of carbon dioxide (CO_2). To convert the amount of carbon into carbon dioxide, we need to multiply the amount of carbon by 3.667 (the ratio of CO_2/C molecular masses, which corresponds to the ratio: 44/12 or 3.67) (IPCC, 2006; Mcghee et al., 2016). Using this coefficient, the amount of CO_2 sequestered in the study area is presented in the following table:

Table26 :: Quantity of carbon dioxide (CO_2) stored at the sites studied

Website	Zone	Total stock (t CO_2/ha)
Admine	Central	238,33
	Tampon	186,01
	Transition	115,97
Ouameslakht	Central	164,10
	Tampon	116,76
	Transition	72,08

Conclusion

Comparison of carbon stocks in the reservoirs between the different zones in the 2 sites, shows that the total carbon stock is relatively greater in the central zones of both sites, however statistical tests have shown that the zoning effect is significant only in the Ouameslakht site, highlighting the effectiveness of set-aside as a management strategy to protect ecosystems.

It should be pointed out that core areas sequester more carbon than buffer and transition zones. This is linked to the choice of core areas by managers, since these areas present a low-disturbance wilderness aspect.

Chapter 2: Study of land use dynamics in the RBA between 1998 and 2021

Introduction

The next chapter traces the evolution of land use dynamics by analyzing the change in land use classes according to the area criterion. Two years have been chosen for this purpose. This choice is not random, but is intended to provide the best possible account of the impact of the argan tree's reservation since 1998 on land-use dynamics and soil carbon stock in the study area. The study period runs from 1998 to 2021.

To do this, we used the Trends.Earth tool as an international conservation platform to monitor land changes using Earth observations in a desktop system

2.1. Assessment of land use trends in the RBA

Analysis of trend data (Tables 27 and 28) shows that rangeland and bare soil occupations have experienced a regressive trend over 23 years, with rangeland and bare soil at -3.81% and -10.46% respectively. This represents an annual rate of about -3,111 ha and -1,396 ha for the two occupations respectively. The rangelands that changed destiny were converted into unserviced land and other land (1873 ha) and cultivated land and forests (104,585 ha). Bare soil and other land that changed class were converted to rangeland (28,796 ha) and cultivated land (4,118 ha). Although the balance of forest land and cultivated land is positive, they have been transferred to rangelands (5,124 ha and 99,461 ha respectively).

Table27 : Evolution of land cover in the biosphere reserve between 1998 and 2021

Land use	Area in 1998 (ha)	Area in 2021 (ha)	Dynamics between 1998 and 2021 (ha)	Percentage (%)
The forest	139 196	152 996	13 800	9,91%

Course	1 879 473	1 807 917	-71 556	**-3,81%**
Agriculture	637 555	722 677	85 123	13,35%
Human settlements	19 291	23 859	4568	23,68%
Bare ground	306 942	274 825	-32 117	**-10,46%**
Wetlands and water bodies	4628	4809	476 360	3.94
Total :	2 987085	2 987085	0	

Table28 : Matrix of land use evolution (conversion) (ha) in the Arganeraie biosphere reserve between 1998 and 2021

1998 \ 2021	The forest	Course	Agriculture	Human settlements	Bare ground	Wetlands	*Total :*
The forest	139 008	27	161	0	0	0	139 196
Course	5 124	1 773 015	99 461	199	1 546	129	1 879 473
Agriculture	88,63	60,79	6 189,37	35,30	1,28	0,16	637 555
Human settlements	0	0	0	19 291	0	0	19 291
Bare ground	0	28 796	4 118	823	273 151	54	306 942
Wetlands	0	0	0	16	0	4 612	4 628
Total :	152 996	1 807 917	722 677	23 859	274 825	4 810	2 987 085

Overall land use trends in the RBA between 1998 and 2021 show that the majority of land has been stable (94.67%). Improved land cover represents 4.91%. On the other hand, land with degraded cover represents less than 0.42% of the surface (Table 29). The main features of this regressive transformation can be summarized as follows:

✓ Transformed forest land has been converted to rangeland;

✓ The rangelands that were transformed were converted to bare soil;

Grazing is the most important factor in the degradation of forest and cultivated land in the RBA.

Table29 : Summary of land cover changes between 1998 and 2021

Type of land cover evolution	Surface (ha)	Percentage (%)
Land with improved cover	146 362	4,91%
Land with stable cover	2 823 418	94,67%
Land with degraded cover	12 490	0,42%

Figures 23, 24 and 25 show the land use maps of the Arganeraie Biosphere Reserve for 1998 and 2021, as well as the land degradation map.

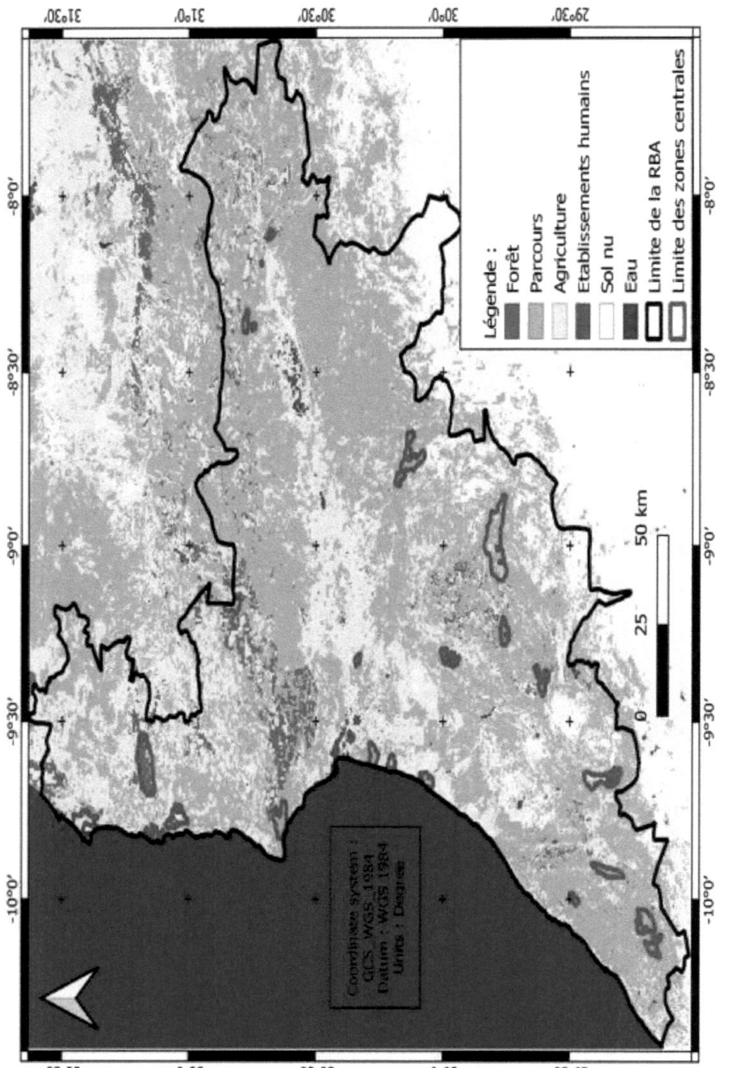

Figure23 : Land use map of the RBA in 1998

82

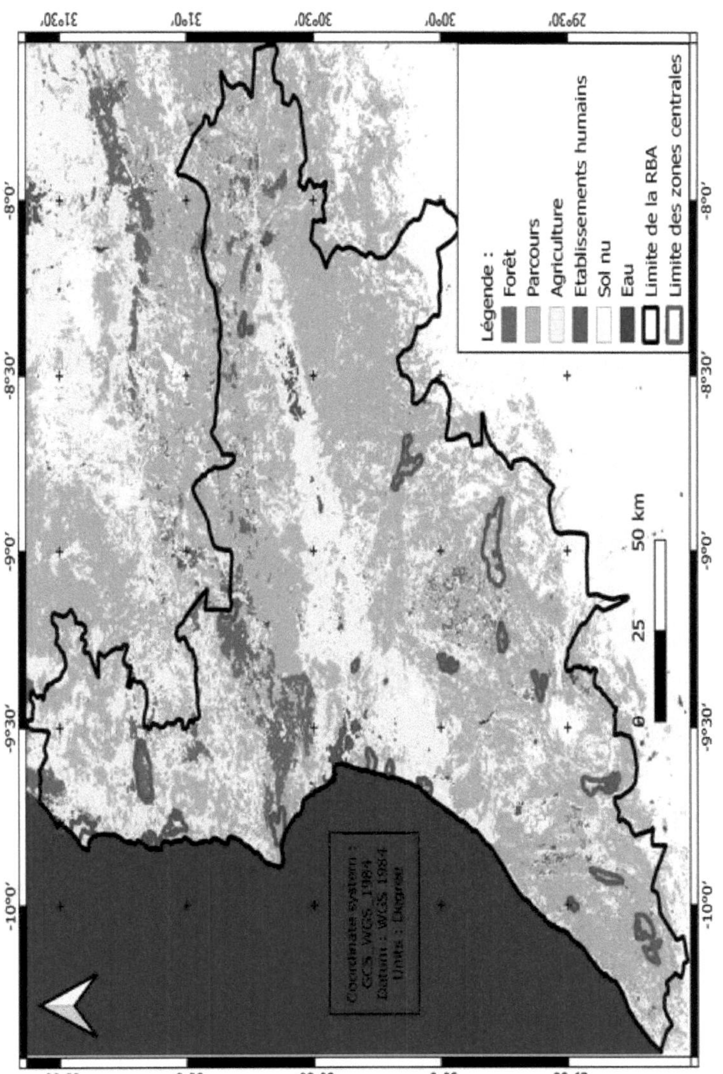

Figure24 : Land use map of the RBA in 2021

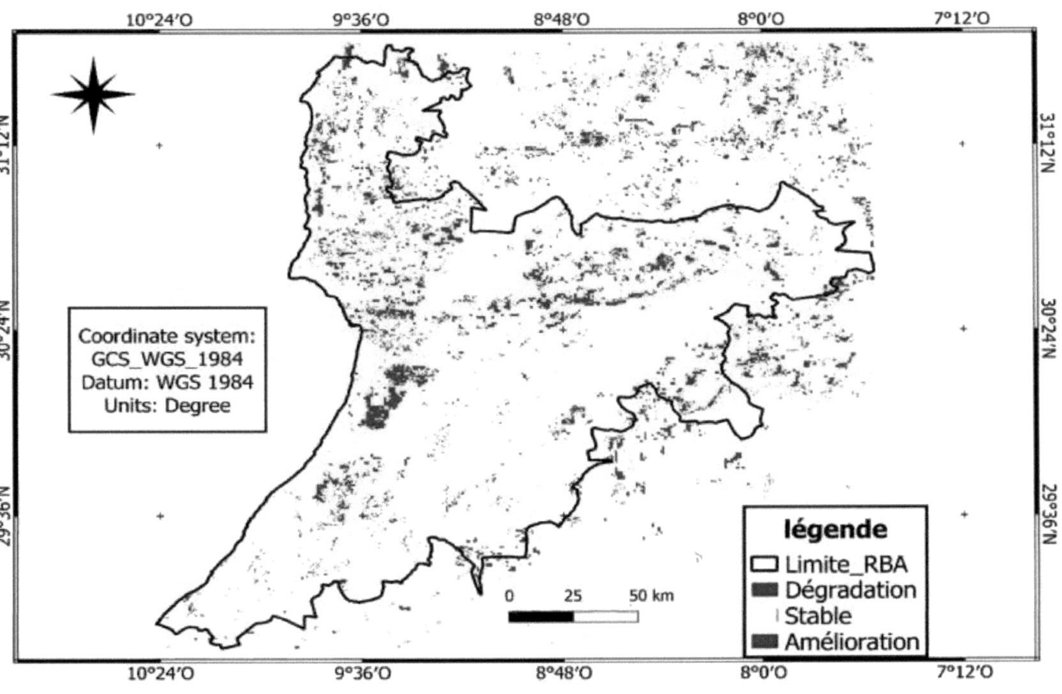

Figure25 : Map of land use degradation in the RBA between 1998 and 2021

2.2.Assessment of land use trends in the central zone of the RBA

Analysis of trend data (Tables 30 and 31) shows that 4 land occupations have experienced a regressive trend over 23 years: forest, parkland, wetlands and bare soil, with respectively -9.33%, -10.27%, -0.35%, -25.27%. This represents, for the four occupations respectively, an annual rate of the order of -4.43 ha, -1,687 ha, -0.23 ha, -709.25 ha. Forest land that has changed use has been converted to cropland (102 ha). Rangelands that have changed use have been converted to farmland (52,720 ha). Bare soil and other land that changed class were converted to parkland (13,471 ha) and cultivated land (2,240 ha).

Table30 : Evolution of terrestrial coverage in the central zone of the RBA between 1998 and 2021

Land use	Area in 1998 (ha)	Area in 2021 (ha)	Dynamics between 1998 and 2021 (ha)	Percentage (%)
The forest	1091	989	-102	**-9,33%**
Course	377 880	339 061	-38 819	**-10,27%**
Agriculture	284 230	335 405	51 176	18,01%
Human settlements	16 223	20 287	4064	25,05%
Bare ground	64 558	48 245	-16 313	**-25,27%**
Wetlands and water bodies	1514	1509	-5	**-0,35%**
Total :	745 497	745 497	0	

Table31 : Matrix of land use change (conversion) (ha) in the central zone of the RBA between 1998 and 2021

1998 \ 2021	The forest	Course	Agriculture	Human settlements	Bare ground	Wetlands	*Total :*
The forest	989	0	102	0	0	0	1091

Course	0	324 929	52 720	194	38	0	377 880
Agriculture	0	661	280 344	3193	32	0	284 230
Human settlements	0	0	0	16 223	0	0	16 223
Bare ground	0	13 471	2240	672	48 176	0	64 558
Wetlands	0	0	0	5	0	1509	1514
Total :	989	339 061	335 405	20 287	48 245	1509	745 497

Overall trends in land cover in the RBA between 1998 and 2021 show that 0.66% of land has been degraded. Improved land cover represents 9.2%. However, land with stable cover represents 90.14% of the surface area (Table 32). This transformation can be described as follows:

- ✓ Forest and rangelands have been converted to agriculture;
- ✓ Bare soil has been transformed into human settlements;

Agriculture is the most important factor in the degradation of forest land and rangelands in the Souss et Dir plain.

Table32 : Summary of land cover changes in the central zone of the RBA, between 1998 and 2021

Type of land cover evolution	Surface (ha)	Percentage (%)
Land with improved cover	68 430	9,20%
Land with stable cover	670 670	90,14%
Land with degraded cover	4 890	0,66%

Figures 26, 27 and 28 show the land use maps of the Arganeraie Biosphere Reserve for 1998 and 2021, as well as the land degradation map.

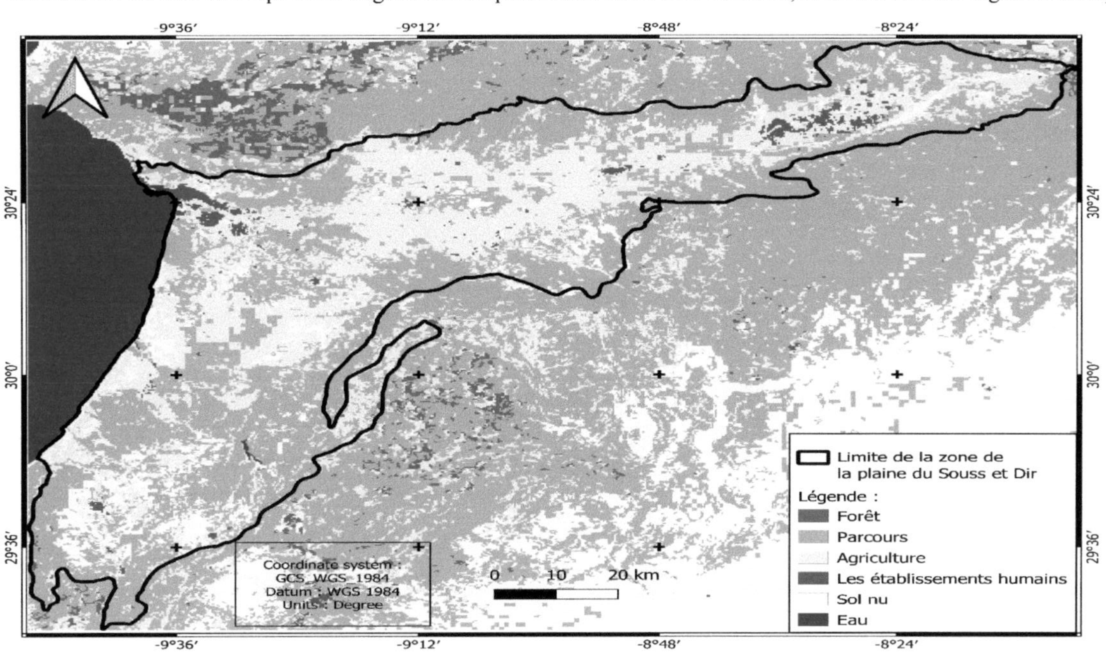

Figure26 : Land use map of the central zone of the RBA in 1998

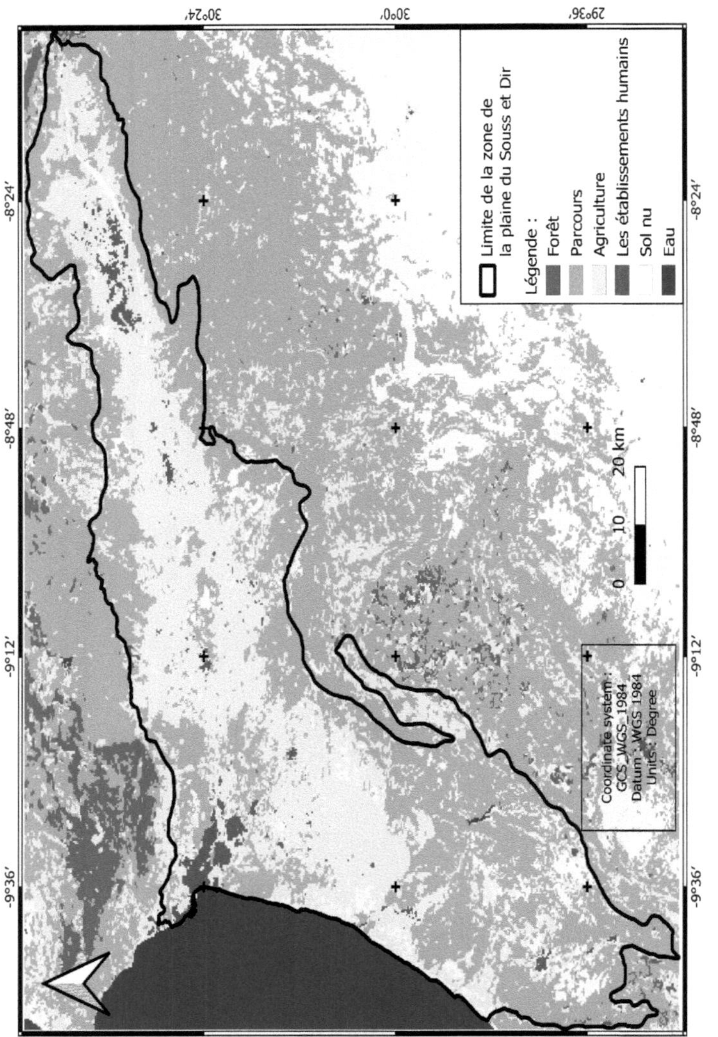

Figure27 : Land use map of the central zone of the RBA in 2021

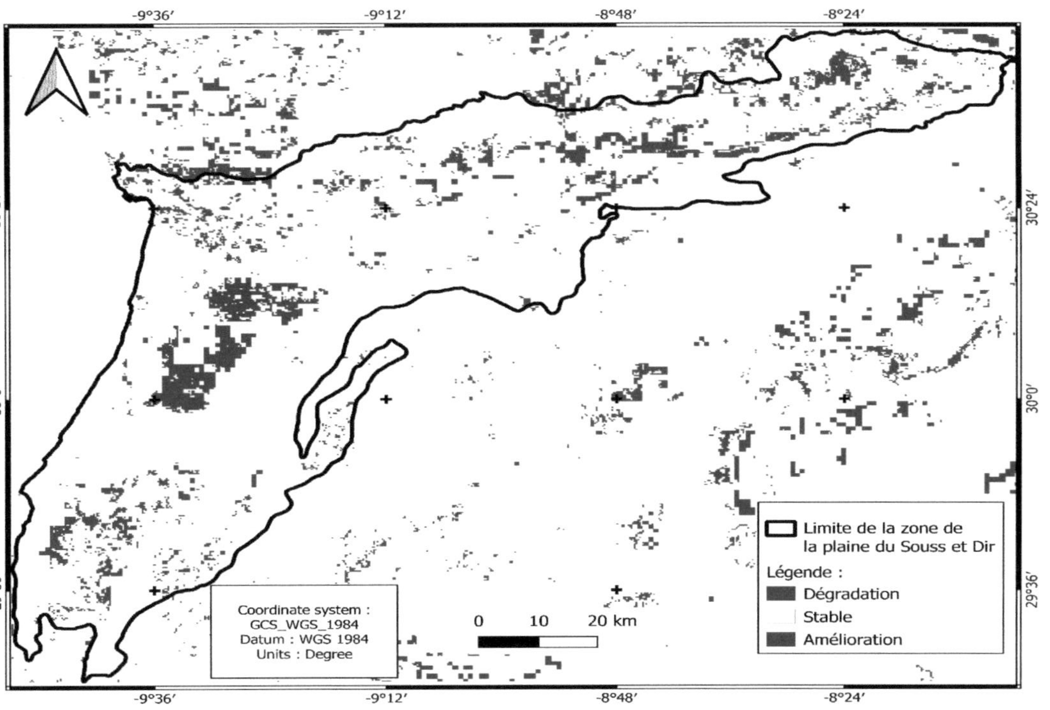

Figure28 : Map of land use degradation in the central zone of the RBA between 1998 and 2021

2.3.Assessment of soil organic carbon stock trends in the central zone of the RBA

In terms of the total surface area of the central zone of the RBA, the COS stock is stable. Between 1998 and 2021, despite the dynamics of land use surfaces, 97.29% of the surface area of the Souss et Dir plain would have a stable COS stock. Areas with an improvement in COS stock would represent 0.61%, and those with a deterioration would account for 2.09% (Table 33).

Table33 : Summary of changes in soil organic carbon between 1998 and 2021 in the central zone of the RBA.

Carbon stock evolution	Surface area (ha)	Percentage of total surface area (%)
Land area with COS improvement	4 560	0,61
Land area with stable COS	723 846	97,29
Land area with COS degradation	15 580	2,
Total surface area	743 990	100

Numerical results for soil organic carbon stock (SCOS) between the year the RBA was created (1998) and the final year 2021 (23 years) (Table 34 and Figures 29 and 30) show that forest, rangeland and bare soil occupations and other land show a net loss in the SCOS stock balance of -11.17%, 12.93% and -14.36% respectively. The fall in the SOC stock balance over this period is also due to the fall in carbon rates per unit area (t C/ha) between 1998 and 2021. For forests, the carbon rate fell from 36.54 t C/ha to 35.79 t C/ha, while for rangelands, the carbon rate fell from 23.07 t C/ha to 22.39 t/ha. This could be explained by changes in rangeland structure. In this respect, overgrazing could be responsible in rangelands and cultivation in bare soils and other lands.

At forest level, our results are in line with the numerical results for soil organic carbon stock (SCOS) in 2021, with an average stock of 35 t C/ha.

Table34 : Changes in soil carbon stock (SCC) by land use in the central RBA from 1998 to 2021

land use	Stock COS 1998 (tons/ha)	Stock COS 2021 (tons/ha)	Surface area 1998 (ha)	Surface area 2021 (ha)	Stock COS 1998 (tons)	Stock COS 2021 (tons)	Change COS stock (tons)	Change Stock COS (%)
Forest	36,54	35,79	1091	9890	39 863,47	35 410,65	-4 452,82	**-11,17%**
Course	23,07	22,39	377 880	339 061	8 717 950,82	7 590 343,11	-1 127 607,71	**-12,93%**
Agriculture	30,35	30,25	284 230	335 405	8 626 823,11	10 146 855,50	1 520 032,38	17,62%
Human settlements	40,31	40,31	16 223	20 281	653 884,69	817 450,80	163 566,12	25,01%
Bare ground	19,77	22,66	64 558	48 245	1 276 480,39	1 093 151,16	-183 329,24	**-14,36%**
Total			743 990	743 990	19 315 002	19 683 211	368 208	

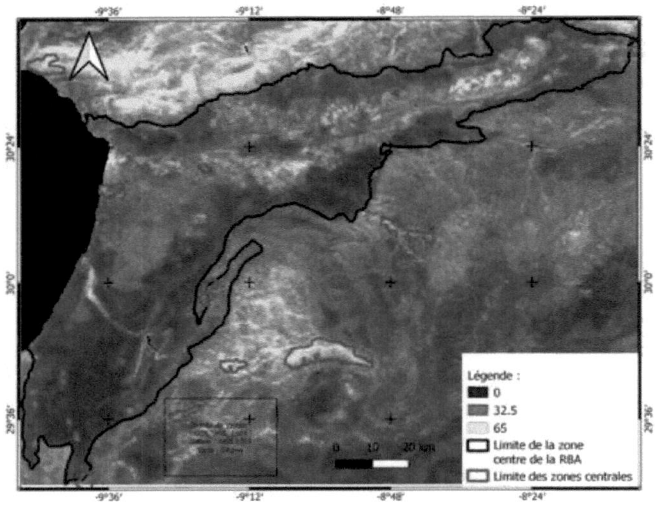

Figure29 : Soil organic carbon stock in the central zone of the RBA in 1998

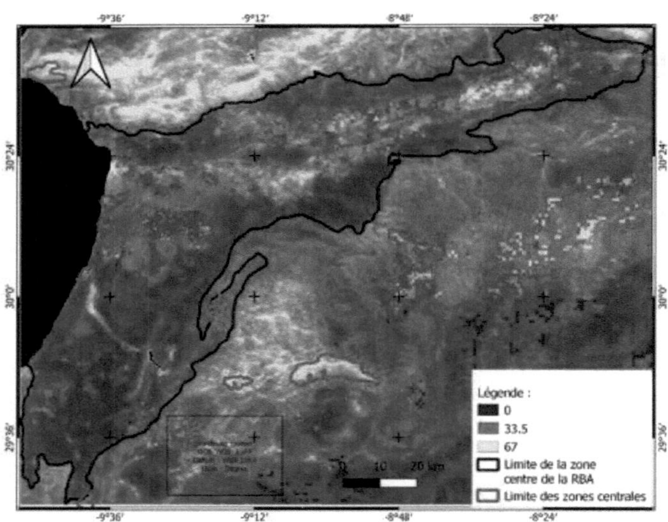

Figure30 : Soil organic carbon stock in the central zone of the ASR in 2021

2.4. Spatio-temporal monitoring of land use dynamics in the central areas of the Souss plain and Dir : Admine and Ouameslakht

Analysis of spatial dynamics at the two power stations was based on a study of changes in the surface area of land-use units. The results of this analysis were obtained in 3 phases:

- The 1st analyzes the land-use situation in the study area at each of the dates 1998 and 2021;

- The 2nd generated land use maps for the dates 1998 and 2021 and presented the areas of the land use classes as well as the coverage rate of each class for each date;

- The 3rd is to detect the change recorded between the two dates and present the results spatially on the change map.

2.4.1.   Land use dynamics in the Admine core zone between 1998 and 2021
    2.4.1.1. Situation in the Admine core zone in 1998

The classification allowed us to distinguish 2 classes (Figure 31):

- Forest,
- Course.

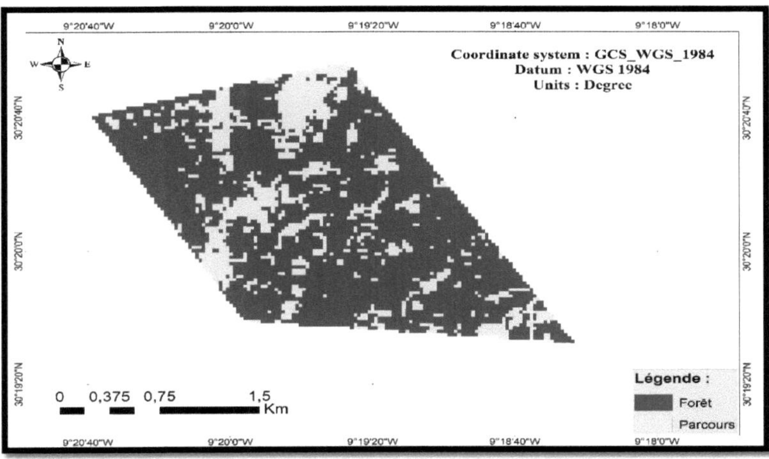

Figure31 : Land use map of the Admine core area in 1998

The land use situation is shown in table 35 and figure 32. Calculations show that the largest area is occupied by forest, with a surface area of 399.73 ha, followed by rangeland, which covers an area of 124.27 ha. These areas show that, in 1998, most of the land in the study area was used for forestry

Table35 : Surface area and coverage rate of land use units in the Admine study area for 1998

Class	Area (ha)	Coverage rate (%)
Forest	399,73	76,3
Course	124,27	23,7

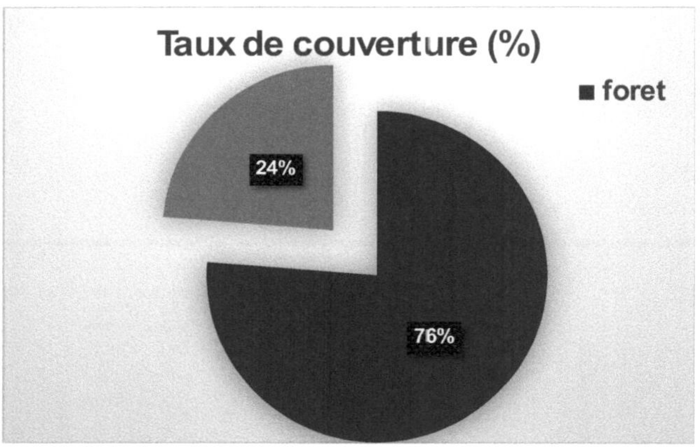

Figure32 : Land use classes in Admine in 1998 in percentages

The classification allowed us to distinguish 2 classes (Figure 33):

- Forest,
- Course.

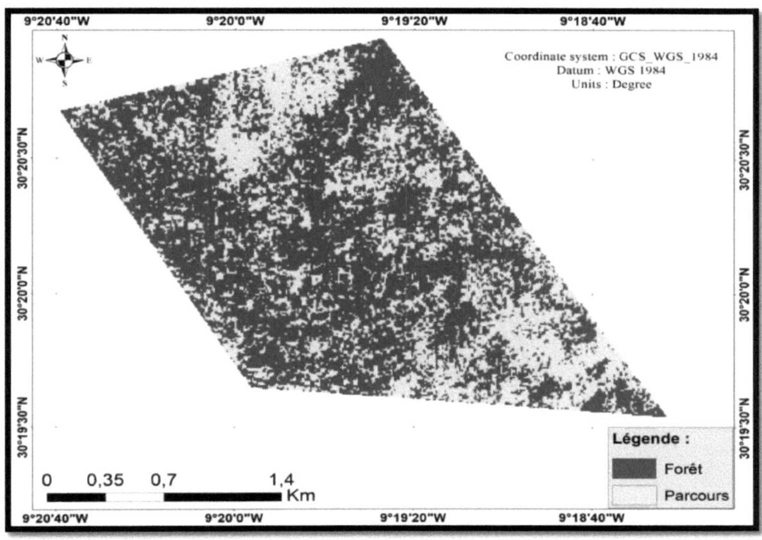

Figure33 :Land-use map of the Admine core area in 2021

The land use situation is shown in Table 36 and Figure 34. According to the statistics generated, the largest areas are occupied by forest and rangeland, with 326.44 and 197.56 hectares respectively

Table36 : Surface area and coverage rate of land-use units in the Admine study area for the year 2021

Class	Area (ha)	Coverage rate (%)
Forest	326,44	62,3
Course	197,56	37,7

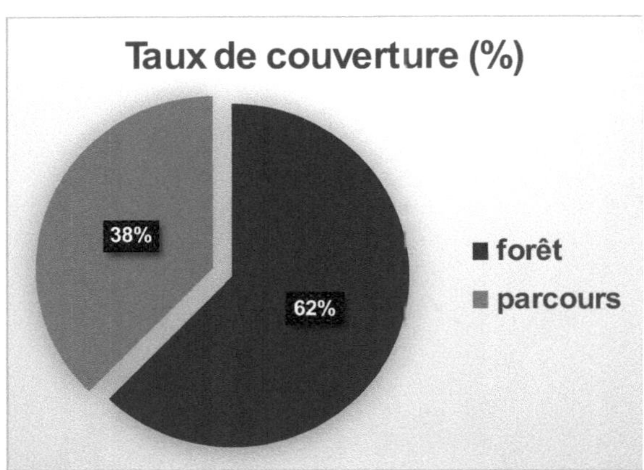

Figure34 : Land-use classes in Admine in 2021 in percent

2.4.1.2.Interpreting classification

The analysis of spatial dynamics in the Admine study area is based on an analysis of changes in the surface area of land-use units over a 23-year period (1998 to 2021). The aim of this study is to present the evolution over time of two classes: forest and rangeland.

Table37 : Evolution of land use units between 1998 and 2021

Class	Area (ha) in 1998	Area (ha) in 2021	Growth between 1986 and 1996 (ha)	Average annual rate (ha)
Forest	399,73	326,44	-73,29	-3,19
Course	124,27	197,56	73,29	3,19

Analysis of Table 37 shows the changes in the study area in terms of area gain and loss for each land use over the period 1998-2021. The gain in rangeland area at the expense of forest land shows that forest degradation is beginning to take place in an area designated for biodiversity conservation by prohibiting all human activity.

✓ The forest regressed at an average annual rate of **-3.19** ha/year;

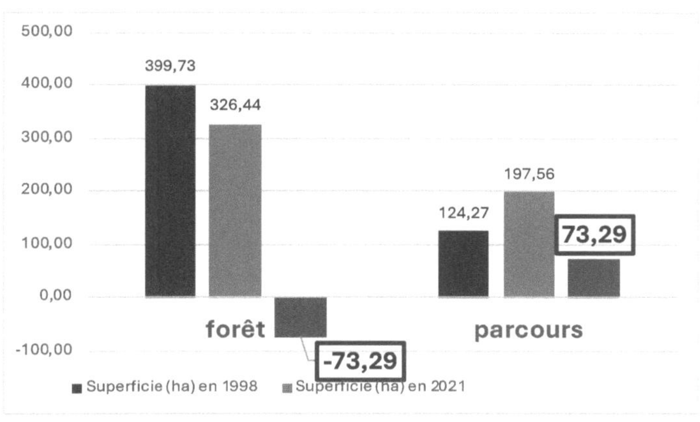

Figure35 : Evolution of land use in Admine between 1998 and 2021

Analysis of figures 35 and 36 shows the changes and evolution of land use units in the study area between 1998 and 2021, represented by the significant regression of the forest class in favor of rangelands.

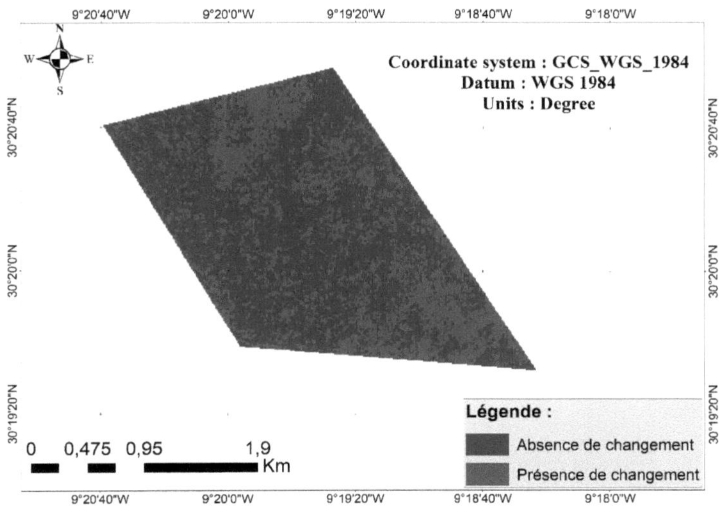

Figure36 : Map of land use changes in the Admine core area between 1998 and 2021

2.4.2. Land use dynamics in the central Ouameslakht zone between 1998 and
2021

   2.4.2.1.Situation of the central Ouameslakht zone in 1998

The classification distinguishes 2 classes:

✓ Forest,

✓ Course.

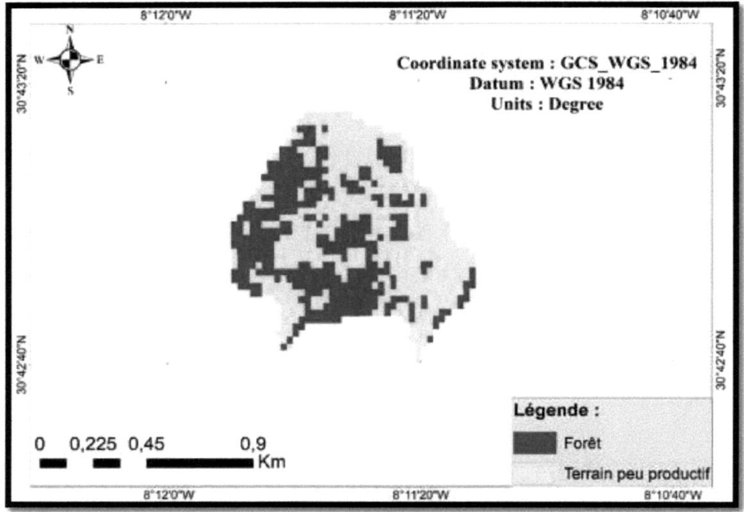

Figure37 : Land use map of the central Ouameslakht zone in 1998

The land use situation is shown in Table 38 and Figures 37 and 38. According to the
statistics generated, the Ouameslakht core zone is divided between forest and
unproductive land, with areas of 30.35 and 41.65 hectares respectively. It should be
noted that the area occupied by forest represents 42.15% of the total area of the core
zone.

**Table38 : Area and coverage rate of land use units in the Ouameslakht study
area for 1998**

Class	Area (ha)	Coverage rate (%)
Forest	30,35	42,15

Unproductive land	41,65	57,65

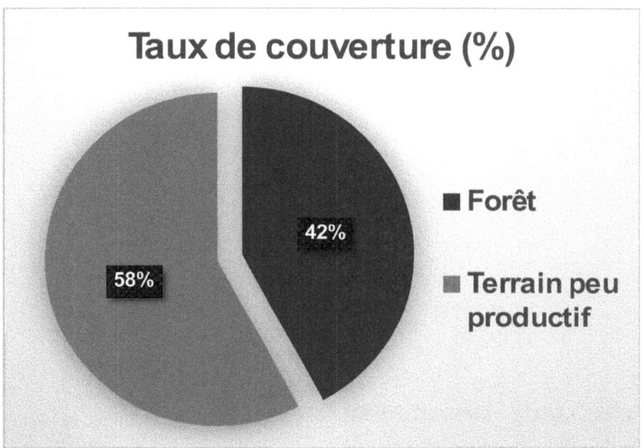

Figure38 : Land-use classes in Ouameslakht in 1998 in percentage

2.4.2.2.Situation of the central Ouameslakht zone in 2021

The classification distinguishes 2 classes:

- Forest,
- Unproductive land.

Ouameslakht's situation is illustrated in the figure below:

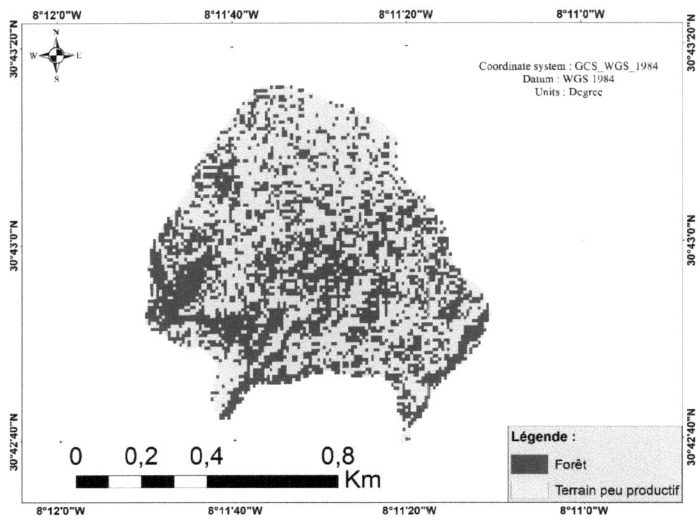

Figure39 : Land-use map of the central Ouameslakht zone in 2021

The classification distinguishes 2 classes of land use:

- Drill,
- Unproductive land.

The land use situation is shown in table 39 and figures 39 and 40.and According to the statistics generated, the largest areas are occupied by forest and unproductive land, with a surface area of 31.14 and 40.86 hectares respectively. Unproductive land occupies 40.86 Ha with a percentage of 57%.

Table39 : Area and coverage rate of Ouameslakht study area land-use units for the year 2021

Class	Area (ha)	Coverage rate (%)
Forest	31,14	43,25
Unproductive land	40,86	56,75

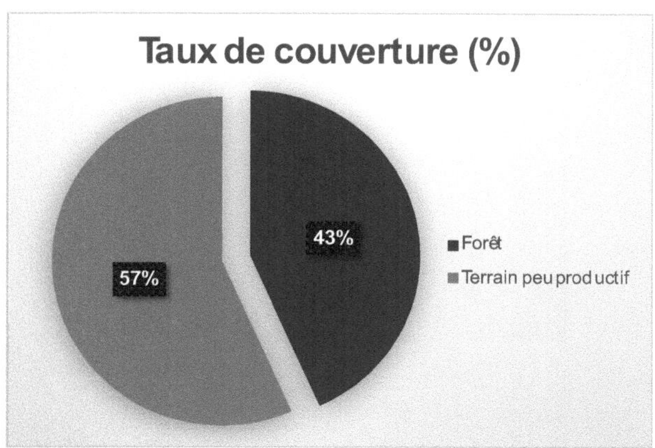

Figure40 : Land use classes in Ouameslakht in 2021 in percent

2.4.2.3.Interpreting classification

The analysis of spatial dynamics in the Ouamselakht study area is based on an analysis of changes in the surface area of land-use units over a 23-year period (1998 to 2021). The aim of this study is to present the evolution over time of two classes: forest and low-productivity land.

Table40 : Evolution of land use units between 1998 and 2021

Class	Area (ha) in 1998	Area (ha) in 2021	Growth between 1986 and 1996 (ha)	Average annual rate (ha)
Forest	30,35	31,14	0,79	0,03
Unproduct ive land	41,65	40,86	-0,79	-0,03

The analysis in Table 40 shows the changes in the study area in terms of area gains and losses for each land use over the period 1998-2021.

➢ The forest layer grew at an average annual rate of+ **0.03 ha/year**;
➢ The stratum of less productive land regressed at an average annual rate of - **0.03 ha/year**;

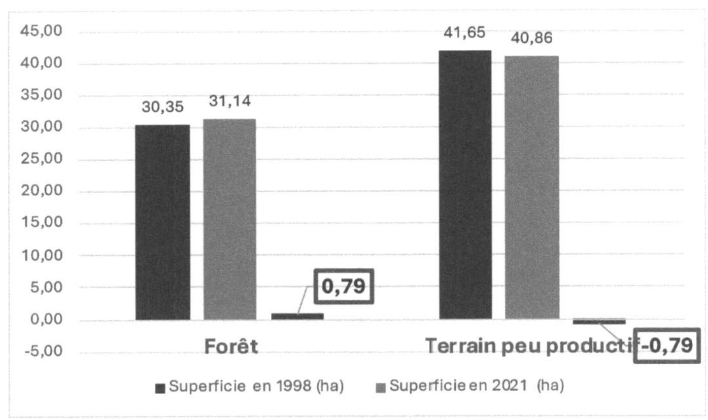

Figure41 : Evolution of land use in Ouameslakht between 1998 and 2021

Analysis of figures 41 and 42 shows the changes in the study area and the evolution of land use units in the study area between 1998 and 2019, represented by the regression of the low-productivity land class in favor of forest. This leads us to assume that the fencing installed in the central zone would have a positive effect on the restoration of the argan tree ecosystem.

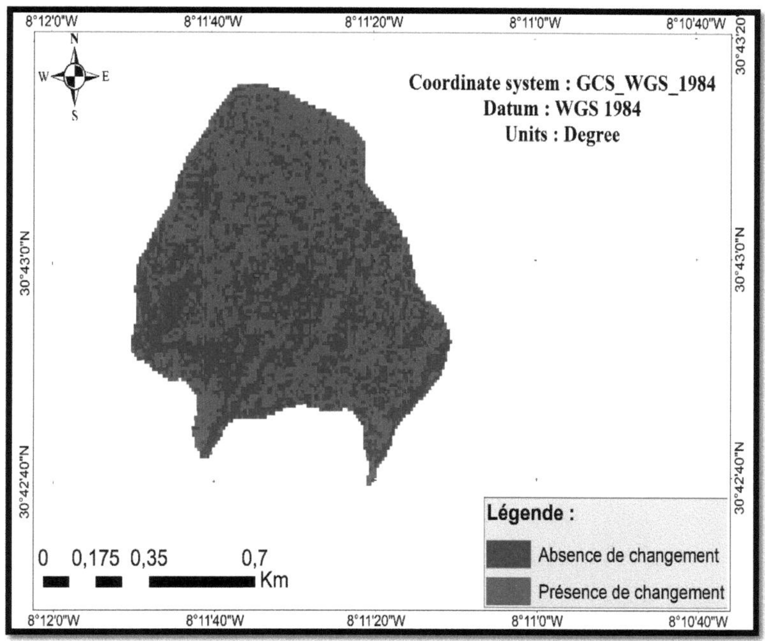

Figure42 : Map of land use changes in the central Ouamslakht zone between 1998 and 2021

Conclusion

Trends.Earth has made it possible to assess land cover by taking stock of the state of land degradation in the RBA and the central zone of the RBA.

According to the results provided by Trends.Earth, the RBA has undergone changes marked essentially by a stabilization of its land cover over 94.67% of its surface area. Degraded occupations in terms of cover represent 0.42%, while land with improved cover accounts for 4.91% of the reserve's total area.

As for the central zone of the RBA, the majority of areas showed stable land cover, with a percentage of 90.14%. Areas with an improvement in land cover represented 9.20% of the total area, compared with virtually no deterioration (0.66%).

A study of satellite images of the central area of Admine from 1998 and 2021 shows a developed spatial dynamic. As needs increase, so will activities. This logic is reflected in the increase in grazing land at the expense of forest.

However, the Ouameslakht zone has seen an increase in forest area, and establishment of set-asides is proving very necessary to guarantee the sustainability of argan stands, while at the same time protecting them from human intervention.

General conclusion and recommendations

In order to reduce the impact of traditional local socio-economic activities on the Arganeraie, the French government has recognized it as an Arganeraie Biosphere Reserve (ABR). It was declared the first biosphere reserve by UNESCO in 1998.

The aim of our study is to assess the effect of the argan tree reserve (RBA) on the carbon sequestration potential of argan tree ecosystems, particularly in the biogeographical region of the Souss and Dir plains. 2 sites have been selected (Admine and Ouameslakht) and three zones have been identified in each: the core, buffer and transition zones.

Initially, this work focused on estimating carbon stocks in the various reservoirs (above-ground biomass, below-ground biomass, necromass and soil) in the 2 sites, and based on random stratified sampling, we took 10 replicates per zone, i.e. 60 plots in total. Using biomass models, we were able to estimate above-ground biomass (woody and foliar). Dry mass conversion enabled us to estimate the carbon stock in above-ground biomass and dead wood. Finally, soil carbon stock was calculated by taking into account bulk density, soil organic carbon concentration and soil depth over 30 cm. This enabled us to study the effect of zoning on carbon stock.

The quantification of carbon in the various reservoirs shows that :

- ✓ Soil is the main reservoir of organic carbon in each zone, ranging from 17.46 to 75.94 t C. ha^{-1}. In fact, soil holds around 85% of the total organic carbon stock, more than two-thirds of the total carbon stock;
- ✓ Above-ground biomass is the second largest carbon reservoir (11% of total carbon), ranging from 1.46 to 6.05 t C ha^{-1}, while woody biomass contributes 94% of the above-ground carbon stock;
- ✓ Underground biomass is the third largest carbon reservoir (3% of total carbon), ranging from 0.29 to 1.21 t C ha^{-1} ;
- ✓ Necromass comes last with 1% of total carbon, ranging from 0.54 to 0.87 t C ha^{-1}, while litter contains between 0.11 and 0.46 t C ha^{-1} of organic carbon, giving an average carbon stock in litter of around 0.27 t C ha^{-1}, and a contribution of 43% to the total stock of necromass. Dead wood stores 0.35 t

C ha^{-1}, contributing 57%. These rather low values can also be explained by the pressure to which the forest is subjected, including livestock grazing and the collection of dead wood and firewood by the local population.

The results of this study also show that the comparison of carbon stocks in reservoirs between different zones in the 2 sites, shows that the total carbon stock is relatively greater in the central zones of both sites, however statistical tests have shown that the effect of zoning is significant only in the Ouameslakht site, highlighting the effectiveness of fencing as management tool to protect ecosystems.

Secondly, we studied the dynamics of land use in the Admine and Ouameslakht core zones between 1998 and 2021. On the one hand, supervised classification enabled us to distinguish the classes: Forest, Rangeland and Less productive land, while diachronic analysis enabled us to highlight the various changes in land use.

In Admine, the forest has undergone a significant regression in favor of grazing land, at around 73 ha, due to overgrazing in particular and anthropic pressure in general. At Ouameslakht, the argan forest has increased by 0.80 ha, thanks to the introduction of a defensive system prohibiting cultivation and transhumant movements.

The carbon stock estimation method used in this work, based on allometric equations, remains acceptable from a methodological point of view, and the results obtained allow a suitable level of precision, despite the fact that these equations are not developed at the scale of each site, and the absence of shrub and herbaceous strata meant that it was not possible to calculate the quantity of carbon sequestered in the entire ecosystem.

In light of the above, we recommend the following:

- Reinforce the protection of core areas against deforestation and degradation factors (expansion of agricultural land, overgrazing) through the installation of fencing (physical and/or social), which can be combined with active afforestation activities. This may include the development of practices such as assisted natural regeneration, reforestation or agroforestry. This will be the case, in particular, in areas that have been degraded, but also on the periphery of core areas and in their buffer zones;

- Analyze changes in land use in different core areas and their corresponding buffer and transition zones to estimate the rate of conversion from forest to non-forest land;
- Estimating carbon stock in other Moroccan forest ecosystems (tetraclinaie and introduced species);
- Evaluate the effect of stocking other species on the carbon sequestration potential of their ecosystems.

References

Aalde H., Gonzalez P., Gytarsky M., Krug T., Kurz W.A., 2006. Generic methodologies applicable to multiple land-use categories. 2006 IPCC guidelines for national greenhouse gas inventories. worldagroforestry.org, 226 p.

Aamou K., 2013. Estimation de la phytomasse foliaire des futaies adultes de l'Arganier et sa   forage value dans le versant sud d'Amsitten (Commune Rurale d'Imgrad) Mémoire de 3 -ème cycle, ENFI, Salé. 49 p.

Achhal A., 1986. Etude phytosociologique et dendrométrique des écosystèmes forestiers du bassin versant du N'fis (Haut Atlas). Thèse de Doct. Es-Science Univ. Aix, Marseille, 204p.annexes 19 p.

Aguer A., 2015. SOLAG_couverts_vegetaux_et_carbone. Chambre d'agriculture des pays de la loire. N°8 du 06 octobre 2015, 14 p.

Arrouays D., Balesdent J., Germon J.C., Jayet P.A., Soussana J.F., Stengel P., 2002. Stocker du carbone dans les sols agricoles de France, Expertise scientifique collective. Ed INRA, 332 p.

Arrouays D., W. Deslais, J. Daroussin, J. Balesdent, J.L Dupouey, C. Nys, V. Badeau and S. Belkacem, 1999. Stocks de carbone dans les sols de France : quelles estimations ? CRAAF, 85 (6), pp. 278-292.

Askri Z., 2008. Contribution à l'évaluation de la séquestration du carbone par la suberaie et ses formations de dégradation (cas de la kroumirie au nord-ouest de la Tunisie. Mémoire de 3 -ème cycle, ENFI, Salé. 36 p.

Belghazi B., 1990. Etude de l'écologie et la productivité du pin maritime (Pinus pinaster var. Maghrebiana) en peuplements artificiels au Nord du Maroc, Thèse de Doctorat ès-sciences Agronomiques, IAV.Hassan II, Rabat, 189 p.

Belghazi T., Ponette Q., Jonard M., Belghazi B., 2016. Woody and foliar biomass of argan coppices in the Haha Plateau (Morocco), pp. 159-166.

Belghazi T., Ouswati S., El Messoussi S., Chakib E., 2017. Sequestration of organic carbon at an argan coppice in the Haha Plateau, pp. 211-214.

Benabid A., 2000. Flora and ecosystems of Morocco: Biodiversity assessment and preservation. Ibis Press, 360 p.

Benchekroun F., Buttoud G., 1989. L'arganeraie dans l'économie rurale du sud-ouest marocain. Forêt méditerranéenne t. XI, no. 2, pp. 127-133.

Bendaanoun M., 1991. Contribution à l'étude écologique de la végétation halophile, halohygrophile et hygrophile des estuaires, lagunes, deltas et sebkhas du littoral atlantique et méditérranéen et du domaine continental du Maroc : Analyse climatique, pédologique et chimique, phytoécologique, phytosociologique et phytogéographique. Perspectives for management, planning and development. Doc. Etat. Es-sciences Naturelles Université Aix Marseille III, Marseille, 580 p.

Boudy P., 1950. Economie forestière nord-africaine. Tome II, morphologie et traitement des essences forestières. Ed. Larousse, Paris, 525 p.

Boudy P., 1952. Guide du forestier de l'Afrique du nord. Maison Rustique, Paris, pp. 121-120.

Boulmane, M., Makhloufi, M., Bouillet, JP, Saint-André, L., Satrani, B., Halim, M., & El Antry, S. (2010). Estimation of organic carbon stock in Moroccan Middle Atlas Quercus ilex. Acta Botanica Gallica, 157, pp. 451-467.

Brown S., 2002. Measuring carbon in forests: current statuts and future challenges. Environmental Pollution 116, pp. 363-372.

Brown S., and Lugo A.E., 1992. Aboveground biomass estimates for tropical moist forests of the brazilian amazon. Interciercia 17, pp. 8-18.

Brown S., Sathaye, J., Cannell M., Kauppi P., 1996. Management of Forests for Mitigation of Greenhouse Gas Emissions. Working Group II, 1995 Second Assessment Report, Chapter 24, Cambridge University Press, pp. 772-797.

Canadell J.G., and Raupach M. R., 2008. Managing forests for climate change mitigation. Science 320, pp. 1456-1457.

C.E.A.E.Q and M.A.P.A.Q. (2003). Determination of organic matter by determination of organic carbon in agricultural soils: modified Walkley-Black method, MA. 1010 - WB 1.0, Ministère de l'Environnement du Québec, 2003, 10 p.

Chenu C., Klumpp K., Bispo A., Angers D., Colnenne C., Metay A., 2014. Storing carbon in agricultural soils: evaluation of action levers for France. Innovations Agronomiques 37, pp. 23-37.

Crowther T., Snoek M., Bradford C., Rowe K,. Todd-Brown N., Sokol W., Wieder J., Carey M., Machmuller M., Lavallee B., Snoek L., 2016. Quantifying global soil carbon losses in response to warming.Nature, pp. 104-110.

Deb S., Bhanu P., Mandal B., Rakshit A., Singh H., 2015. Soil organic carbon: Towards better soil health, productivity and climate change mitigation. Climate change and Environmental Sustainability, 3(1), pp 26-34.

Derrière N., Wurpillot S., Vidal C., 2012. Dead wood in forests. Issue 29 (June 2012) of the national institute of geographic and forestry information (L'IF), 8 p.

Dixon R.K., Brown S., Houghton R.A., Solomon A.M., Trexler M.C., 1994. Carbon pools and fluxes of global forest ecosystems. Science, pp. 185-190.

Dupouey J.L., Pignard G., Badeau V., Thimonier A., Dhôte J.F., Nepveu G., Bergès L., Augusto L., Belkacem S. and Nys C., 1999. Carbon stocks and fluxes in French forests, C.R. Acad. Agric. Fit, 85(6), pp 293-310.

Emberger L., 1960. Traité de botanique systématique. Les végétaux vasculaires. Tome II, Fasc. 2, Ed Masson, pp. 852-855.

Emberger L., 1955. A biogeographical classification of climates Rev. Trav. Lab. Bot. Geol. Zool, Fac Sci, Univ. Montpellier. 43 p.

Emberger L., 1925. The natural domain of the argan tree. Bull.Soc.Bot. Fr, Tome 72; pp. 770- 774.

Emberger L., 1939. Aperçu sur la végétation du Maroc, Commentaire de la carte phytogéographique du Maroc (au 1/1.500.000) Inst. Sci. Chérif, Rabat, 157p.

FAO & ITPS, 2015. Status of the World's Soil Resources, Rome. : s.n, 48 p.

FAO, 2017. Soil Organic Carbon: an invisible wealth. Food and Agriculture Organization of the United Nations, Rome, Italy. pp. 124-131.

FAO, 2020. Techniques for estimating forest biomass for different carbon pools. 75 p.

IPCC, 2007. Climate Change 2007: Synthesis Report. 104 p.

IPCC, 2014. Climate Change 2014. Impacts, Adaptation and Vulnerability: summary for decision-makers. 30 p.

Godron M., 1976. Phytoecological sampling. Note n° 8, CNRS- CEPE Louis Emberger, 23 p.

Godron M ., 1971. Essai sur une approche probabiliste de l'écologie des végétaux. Thèse Doct. ès Sc. Nat. USTL, Montpellier, 247 p.

Halpin P.N., 1997. Global climate change and natural area protection: management responses and research directions. Ecological Applications 7(3), 828 p.

Hairiah, 2001. Methods of simpling carbon stocks above and blew ground . international center for research in agrofrestry. Bogor, Indonesia. 32 p.

HCEFLCD, 2008. Ten-Year Assessment Report of the Arganeraie Biosphere Reserve (ABR). 2008. 40 p.

HCEFLCD., 2010. Moroccan natural ecosystems and climate change: ecological resilience put to the test. 8 p.

Heller N.E., & Zavaleta E.S., 2009. Biodiversity management in the face of climate change: a review of 22 years of recommendations. Biological conservation 142 : pp. 14-32.

I.F.N., 1999. Moroccan National Forest Inventory. Rapport de synthèse, Direction de développement forestier, Rabat, 45 p.

IPCC. 2018. Global Warming of 1.5°C. An IPCC Special Report on the impacts of global warming of 1.5°C.

IPCC, 2006. Guidelines for National Greenhouse Gas Inventories. Agriculture, Forestry and Other Land Use. 67 p.

Keestrea S., Bouma J., Wallinga J., Tittonell P., Smith P., Cerdà A., Montanarella L., Quinton J., Pachepsky Y., Wim H., van der P., Moll G., 2016. The significance of soils and soil science towards realization of the United Nations Sustainable Development Goals. SOIL, 2: pp. 111-128.

Kozlowski T., Kramer P., Pallardyal S., 1991. The Physiological Ecology of Woody Plants, San Diego, California, Academic Press, inc. 657 p.

Landsberg J.J., Kaufmann R., Binkley D., 1991. Evaluating Progress Toward Closed Forest Models Based on Fluxes of Carbon, Water and Nutrients. Tree physiology, 9(1-2), pp. 1-15.

Lescuyer G., Locatelli B. 1999. The role and value of tropical forests in climate change. Climatic. BOIS ET FORETS DESTROPIQUES, 1999, N° 260 (2), pp. 5-17.

Le Quere C., Michael R., Raupach Josep G., Canadell M.,2009. Trends in the sources and sinks of carbon dioxide . Nature Geoscience 2, pp. 831 - 836.

Lewis L., Gabriela L., Bonaventure S., 2009. Increasing carbon storage in intact African tropical forests . Nature 457, pp. 1003 - 1007.

Losi, C.J., Siccama T.G., Condit R., Morales J.E., 2003. Analysis of alternative methods for estimating carbon stock in young tropical plantations, Forest Ecology and Management184, pp. 355-368.

MacDicken K., 1997. A guide to Monitoring Carbon Storage in Foresty and Agroforesty Projects. Forest Carbon Monitoring Program. Winrock International Institute for Agricultural Development. 92 p.

Mahamane I., 2006. Etude de l'impact des transformations des ecosystemes forestiers sur le stockage de carbone dans la biomasse et dans le sol (cas du bv du bouregreg: région d'oulmès).

Maire R., 1939. The argan groves of Beni Snassen. Botanica Notiser pp: 447-484.

Mao R., Zeng D., Lin Hu H., Jun L., 2010. Soil organic carbon and nitrogen stocks in an age-sequence of poplar stands planted on marginal agricultural land in Northeast China. Plant Soil. 332 (1-2), pp. 277-287.

Maxwell L., Victor C., Nigel D., Michael H., Sue S., 2020. Area-based conservation in the twenty-first century. Nature 586: pp. 217-227.

McGhee W., Saigle W., Padonou E.A., Lykke A.M., 2016. Methods for calculating tree biomass and carbon in West Africa. Annales des Sciences Agronomiques, 20, pp. 79-98.

Mhirit O., Et-Tobi M., 2010. Les écosystèmes forestiers face au changement climatique : Situation et perspectives d'adaptation au Maroc. Changement climatique : Impacts sur le Maroc et options d'adaptation globales Rabat : Institut Royal Des Etudes Stratégiques, 260 p.

M'hirit O., El yousfi S.M., Benzyane M., Benchekroun F. & Bendaanoun M., 1998. L'arganier : une espèce fruitière-forestière à multiples usages. Mardaga. Belgium, 145 p.

Mokssit A., 2012. Climate change update in Morocco. In: Environnement E. ZEINOMAHMALAT and A. BENNIS eds. BENNIS eds, Ed. Konrad-Adenauer-Stiftunge, office in Morocco, pp. 35-39.

Moore J.C., 2018. Predicting tipping points in complex environmental systems. Proceedingsof the National Academy of Sciences 115 (4), pp. 635-636.

Marianne R., 2009. Forests and the carbon cycle. 321, pp. 88-92.

NAGGAR M., 2018. La gestion durable de l'arganeraie et les enjeux de lutte contre la désertification, 84 p.

Noumi N.V., Zapfack L., Pelbara P., Awe Djongmo V., Tabue Mbobda R., 2018. Afforestation/ Reforestation Based on Gmelina Arborea (Verbenaceae) in Tropical Africa: Floristic and Structural Analysis, Carbon Storage and Economic Value (Cameroon). Sustainability in Environment 3 : 161 p.

ONERR (l'Observatoire national d'études et recherches sur les risques)., 2022. Morocco: "The worst drought in 30 years". Available on :

https://www.francetvinfo.fr/monde/afrique/economie-africaine/maroc-la-pire-secheresse-depuis-30-ans_4979094.html/. Visited on April 20, 2022 at 22:00.

Oubrahim H., 2015. Estimation and mapping of the carbon stock in the quercus suber ecosystem of western mamora (cantons a and b) and estimation of its nutrient stock, pp. 122-163.

O'Rourke M., Angers A., Holden M., McBratney B., 2015. Soil organic carbon across scales. Global Change Biology, 21, pp. 3561-3574.

Pan Y., Birdsey R.A., Fang J., Houghton R., Kauppi P.E., 2011. A large and persistent carbon sink in the world's forests. Science. 333, pp. 988-993.

Pellerin S., et Bamière L. (scientific pilots), Launay C., Martin R., Schiavo M., Angers D., Augusto L., Balesdent J., Basile-Doelsch I., Bellassen V., Cardinael R., Cécillon L., Ceschia E., Chenu C., Constantin J., Darroussin

J.,., Delacote P., Delame N., Gastal F., Gilbert D., Graux A-I., Guenet B., Houot S., Klumpp K., Letort E., Litrico I., Martin M., Menasseri S., Mézière D., Morvan T., Mosnier C., Roger-Estrade J., Saint-André L., Sierra J., Thérond O., Viaud V., Grateau R., Le Perchec S., Savini I., Réchauchère O., 2019.** Stocker du carbone dans les sols français, Quel potentiel au regard de l'objectif 4 pour 1000 et à quel coût? Synthèse du rapport d'étude, INRA (France), 114 p. regard de l'objectif 4 pour 1000 et à quel coût? Synthesis of the study report, INRA (France), 114 p.

Peltier J.P., 1982. Vegetation of the Oued Souss watershed (Morocco). Univ. Grenoble thesis, 201p + appendices.

Pearson T., Brown S., 2005. Exploring the Carbon Sequestration Potential of Classified Forests in the Republic of Guinea: A Guide to Measuring and Monitoring Carbon in Forests and Grasslands. Winrock International, Arlington, VA, USA. 39 p.

Pregitzer E., Euskirchen S., 2004. Carbon cycling and storage in world forests: biome patterns related to forest age. Global Change Biol ; (10) : 2052 p.

Quenea K., 2004. Structural and dynamic study of refractory lipid and organic fractions in soils from a forest/grassland chronosequence (CESTAS, South-West France). PhD thesis, Université Paris VI, France. 191 p.

Rhoufacha M., Effet de l'âge des peuplements de chêne-liège de la forêt de la Maâmora sur les différents réservoirs du stock du carbone 2021. Mémoire de 3 -ème cycle, ENFI, Salé. 49 p.

Rieuf P., 1962. Les champignons de l'arganier. Les cahiers de la recherche Agronomique, INRA, Rabat, 15 p.

Ruiz-Peinado R., Bravo-Oviedo A., López-Senespleda E., Montero G., 2013. Do thinnings influence biomass and soil carbon stocks in Mediterranean pinewoods? European Journal of Forest Research. 132, pp. 253-262.

Ryan G.M., 1991. A Simple Method for Estimating Gross Carbon Budgets for Vegetation in Forest Ecosystems. In: Tree Physiology, 9 (1-2), p. 255-266.

Scaramuzza, P., Micijevic, E., 2004. Chander, G.: SLC Gap-Filled Products Phase One Methodology. 145 p.

Shaiek O., Loustau D., Garchi S., Bachtobji B., EL aouni M.H., 2010. Allometric estimation of maritime pine biomass in littoral dunes: the case of Rimel forest (Tunisia). Revue Forêt méditerranéenne, XXXI, n° 3, pp. 231-241.

Vigot M., 2012. Le carbone organique des sols cultivés de Poitou-Charentes ; quantification et évolution des stocks. Chambre d'agriculture de PoitouCharentes/RMT Sols et territoires, 24 p.

Waring R.H., and W.H. Schlesinger, 1985. Forest Ecosystems: Concepts and Management, Academis Press, Orlando, Florida, 340 p.

Walkley A., and Black I.A., 1934. An examination of the Degtjareff method for determining soil organic matter and a proposed modification of the chromic acid titration method. Soil Sci. 34: pp. 29-38.

Woomer P.L., Tieszen L.T., Tschakert P., Parton W.T., Touré A., 2001. Landscape Carbon Sampling and Biogeochemical Modeling: A two-week skills development workshop conducted in senegal. SACRED Africa, Nairobi, Kenya, 69 p.

Zaher H., Benjelloun H., Mahamane I., 2019. Effect Oak Ecosystems Degradation on the Carbon Storage in the Southern Mediterranean Forests. Journal of environmental and science. DOI: 10.32474/OAJESS.2019.04.000185.

Appendices

Appendix1 : Inventory sheet by plot and stratum

1. Argan tree dendrometric measurements :

<table>
<tr><td colspan="3">PLACETTE N° :</td><td colspan="4">ZONE :</td></tr>
<tr><td colspan="3">Contact details :</td><td colspan="2">X=</td><td colspan="2">Y=</td></tr>
<tr><td rowspan="2">Tree no.</td><td rowspan="2">Diet</td><td rowspan="2">C $_{1.30}$
(cm)</td><td rowspan="2">H (m)</td><td rowspan="2">Number of
sprigs (per
vine)</td><td colspan="2">Diameter
From the crown</td></tr>
<tr><td>D1</td><td>D2</td></tr>
<tr><td></td><td></td><td></td><td></td><td></td><td></td><td></td></tr>
<tr><td></td><td></td><td></td><td></td><td></td><td></td><td></td></tr>
</table>

2. Measurement of dry matter in necromass

<table>
<tr><td colspan="2">ZONE :</td><td></td></tr>
<tr><td>Contact details :</td><td>X=</td><td>Y=</td></tr>
<tr><td>Stratum</td><td>Fresh weight (g)</td><td>Dry weight (g)</td></tr>
<tr><td>Litter</td><td></td><td></td></tr>
<tr><td>Dead wood</td><td></td><td></td></tr>
</table>

Figure43 : Measuring the height and circumference of an argan tree in the central zone of Ouameslakht.

Figure44 : Litter sampling

Figure45 : Samples of dead wood from different areas

Figure46 : Soil samples from different areas

Figure47 : Drying soil samples to measure bulk density

Appendix3 : Excessive grazing in the Admine core zone

Printed by Books on Demand GmbH, Norderstedt / Germany